AF616337

NOT TO BE REMOVED
FROM THE
WITHDRAWN

BOURNEMOUTH
UNIVERSITY
WITHDRAWN
LIBRARY

3 Forensic Science Progress

Forensic Science Progress

Volume 3

With Contributions by
C. A. Pounds, R. N. Smith

With 19 Figures and 7 Tables

Springer-Verlag
Berlin Heidelberg New York
London Paris Tokyo

Editors-in-Chief

Prof. Dr. A. Maehly
Forensic Science Centre, 21 Divett Place, Adelaide
5000 S.A./Australia

Prof. Dr. R. L. Williams
Director, Metropolitan Police Forensic Science Laboratory,
109 Lambeth Road, London SE1 7LP/England

ISBN-3-540-18447-3 Springer-Verlag Berlin Heidelberg New York
ISBN-0-387-18447-3 Springer-Verlag New York Heidelberg Berlin

The Library of Congress has cataloged this serial publication as follows:
Forensic science progress. — Vol. 1- — Berlin; New York: Springer-Verlag, c1986-
v.: ill.; 25 cm.
Editors: v. 1- A. Maehly, R. L. Williams.
ISSN 0930-1461 = Forensic science progress.
1. Criminal investigation—Periodicals. 2. Criminal investigation—Collected works. I. Maehly, Andreas Christian, 1917-. II. Williams, R. L.
[DNLM: 1. Forensic Medicine—periodicals. W1FO615P]
HV8073.F588 363.2′5′05—dc19 86-640073 AACR 2 MARC-S
Library of Congress [8707]

Typesetting: Hagedorn, Berlin; Offsetprinting: Heenemann, Berlin;
Bookbinding: Lüderitz & Bauer, Berlin
2154/3020-543210

Editorial Board

Editorial

During the years 1962–1965 Interscience Publishers produced a four-volume series called "Methods in Forensic Science". Since then no major effort seems to have been made to review the progress in the rapidly expanding field of forensic science.

Our new series "Forensic Science Progress" represents a serious effort to take up a neglected venture. The series intends to cover all aspects of forensic science but does not include forensic medicine which is well represented in other publications.

The aim of the publisher and the board of editors is to produce contributions of high quality by leading scientists in the field of forensic science. Suggestions for such contributions from the forensic science community at large are of course also very welcome.

The volumes will not be topic-oriented but will give balanced views on various aspects of the science. The editors believe that the forensic worker should be informed about all branches of the science even if he may very well be specialised in one or a few of them. Ideally, contributions should be from 40–80 typewritten pages. Experimental details, except when not published previously, should be covered by citing the appropriate references. Polemic passages should be avoided but this does not exclude objective criticism.

The publisher has tried to choose an editorial board which is representative not only of various topics but also of the various geographical regions of the world.

Editors — Publisher

Table of Contents

Radioimmunoassay of Drugs in Body Fluids in a Forensic Context
R. N. Smith 1

Developments in Fingerprint Visualisation
C. A. Pounds 91

Author Index Volumes 1–3 121

Subject Index 123

Radioimmunoassay of Drugs in Body Fluids in a Forensic Context

R. N. Smith
The Metropolitan Police Forensic Science Laboratory,
109 Lambeth Road, London SE1 7LP, U.K.

Radioimmunoassay (RIA) is a versatile and extremely sensitive technique that can detect low levels of large or small molecules in biological fluids with little or no sample preparation. It is therefore very useful in forensic toxicology for screening body fluids for drugs of abuse. This account describes the theory and practice of RIA with particular reference to the analysis of drugs in body fluids in a forensic context. RIA theory is outlined from basic principles but the inherent assumptions are often inapplicable in practice and so the empirical design of an assay is considered in detail. Particular emphasis is given to the development of assays for drug screening that detect classes of structurally related compounds rather than individual drugs. The preparation of radiolabelled drugs, the synthesis of immunogens for raising anitisera, the production of polyclonal and monoclonal antisera, and methods for separating free and antibody-bound antigens are reviewed. Quality assurance, trouble-shooting and the possible hazards of forensic RIA are discussed, and published RIA methods for drug analysis are tabulated. Many non-isotopic immunoassays have been developed in recent years but are omitted from this account because to date they are less applicable than RIA to samples such as haemolysed blood that are frequently encountered in forensic toxicology. Future progress in forensic drug RIA is likely to be concerned with applying the technique to more compounds, improving the methods for preparing immunogens and radiolabelled drugs, and investigating the use of monoclonal anti-drug antibodies.

List of Symbols 5

1 Introduction 6

2 Theory 7
2.1 Basic Principles 7
2.2 The Equilibrium Constant 8
2.3 Theoretical Immunoassay Model and Optimisation of Assay Conditions 11
2.3.1 Yalow and Berson's Conditions 13
2.3.2 Ekins' Conditions 13
2.3.3 High Antiserum Concentration 14
2.4 Precision, Sensitivity and Accuracy 14
2.4.1 Precision 14
2.4.2 Sensitivity 15
2.4.3 Accuracy 16

3 Practical Design of an Assay 17
3.1 Antiserum Dilution Curve 17
3.2 Calibration Curve 19
3.3 Specific Assay 20

Forensic Science Progress 3

3.4 General Assay 23
3.5 Multiple Drug Assay 24

4 Radioligands 25
4.1 β-Emitting Isotopes 25
4.1.1 β-Labelled Drugs 25
4.1.2 Stability and Storage of β-Labelled Drugs 25
4.1.3 Liquid Scintillation Counting 26
4.2 γ-Emitting Isotopes 29
4.2.1 ^{125}I-Labelled Drugs 29
4.2.2 Oxidative Radioiodination Methods 29
4.2.3 Conjugation Labelling: Homologous and Heterologous Assays . 31
4.2.4 Other Radioiodination Methods 32
4.2.5 Stability and Storage of ^{125}I-Labelled Drugs 32
4.2.6 γ-Counting 33
4.3 Specific Activity 33

5 Antiserum Production 34
5.1 Immunogen Preparation 36
5.1.1 Haptens with Primary Aliphatic Amino Groups 36
5.1.2 Haptens with Secondary Aliphatic Amino Groups 37
5.1.3 Haptens with Aromatic Amino Groups 40
5.1.4 Haptens with Nitro Groups 41
5.1.5 Haptens with Carboxyl Groups 41
5.1.6 Haptens with Aliphatic Hydroxyl Groups 42
5.1.7 Haptens with Phenolic Hydroxyl Groups 43
5.1.8 Haptens with Carbonyl Groups 44
5.1.9 Haptens with Active Hydrogen 44
5.1.10 Haptens with Ester Linkages 45
5.1.11 Haptens with Lactone Rings 45
5.1.12 Haptens with Thiol Groups 46
5.1.13 Miscellaneous Haptens 46
5.2 Purification of Immunogens and Estimation of the Degree of Conjugation 47
5.3 Storage of Purified Immunogens 48
5.4 Immunisation 49
5.5 Collection and Storage of Antiserum 49
5.6 Monoclonal Antibodies 50

6 Separation of the Bound and Free Fractions 51
6.1 Widely-Used Separation Methods 51
6.1.1 Adsorption 51
6.1.2 Fractional Precipitation 52
6.1.3 Second Antibody Precipitation 53
6.1.4 Solid Phase Separations 53
6.2 Less Widely-Used Separation Methods 53
6.2.1 Electrophoresis 53
6.2.2 Gel Permeation Chromatography 54

6.2.3 Ion Exchange . . . 54
6.2.4 Partition between Two Liquid Phases . . . 54
6.2.5 Dialysis . . . 54

7 Quality Assurance . . . 54

8 Trouble-Shooting . . . 55

9 Hazards . . . 56

10 Published RIA Methods for Drug Analysis . . . 57
10.1 Analgesics and Narcotics . . . 57
10.2 Antiasthmatics and Other Drugs Affecting the Respiratory System Including Bronchodilators, Bronchospasm Relaxants and Nasal Decongestants . . . 57
10.3 Antibiotics and Antineoplastics . . . 58
10.4 Anticholinergic Drugs . . . 58
10.5 Anticoagulants . . . 58
10.6 Anticonvulsants . . . 59
10.7 Antidiabetic Drugs . . . 59
10.8 Antidiarrhoeal Drugs . . . 59
10.9 Antiemetic Drugs . . . 59
10.10 Antihistamines . . . 59
10.11 Anti-Inflammatory Drugs and Gout Suppressants . . . 59
10.12 Antimalarial Drugs . . . 60
10.13 Antipsychotic, Hypnotic and Sedative Drugs . . . 60
10.13.1 Barbiturates . . . 60
10.13.2 Benzodiazepines . . . 60
10.13.3 Phenothiazines . . . 60
10.13.4 Tricyclic Antidepressants . . . 61
10.13.5 Other Antipsychotic Drugs . . . 61
10.14 Antithyroid Drugs . . . 61
10.15 Antitubercular Drugs . . . 61
10.16 Anti-ulcer Drugs . . . 61
10.17 Antiviral Drugs . . . 61
10.18 Anti-worm (Anthelmintic) Drugs . . . 61
10.19 Cardiovascular Drugs . . . 62
10.19.1 Adrenergic Blocking Drugs . . . 62
10.19.2 Antiarrythmic Drugs . . . 62
10.19.3 Antihypertensive Drugs . . . 62
10.19.4 Cardiac Depressants . . . 62
10.19.5 Cardiac Glycosides and Aglycones . . . 62
10.20 CNS Stimulants and Psychoactive Drugs . . . 63
10.21 Diagnostic Aids . . . 63
10.22 Diuretic Drugs . . . 63
10.23 Eicosanoids, Leukotrienes and Prostaglandins (Selected References) . 63
10.24 Ergot Derivatives . . . 63

10.25 Ganglionic Stimulant Drugs 64
10.26 Herbicides and Insecticides 64
10.27 Immunomodulating and Immunosuppressive Drugs 64
10.28 Muscle Relaxants 64
10.29 Narcotic Antagonists 64
10.30 Renal Tubular Transport Inhibitors 64
10.31 Sympathomimetic Drugs 65
10.32 Steroids (Selected References) 65
10.33 Tobacco Alkaloids and Metabolites 65
10.34 Toxins 65
10.35 Vasodilators 66
10.36 Vitamins 66

11 References 66

12 Glossary 87

List of Symbols

Ab	antibody
Ab_T	total antibody
Ag	antigen
Ag*	labelled antigen
Ag_T	total antigen
B	concentration of antibody-bound antigen
B*	radioactivity in the bound fraction
B_0^*	radioactivity in the bound fraction in the absence of unlabelled antigen, i.e. as with the zero standard of the calibration curve
F	concentration of free antigen
F*	radioactivity in the free fraction
k_1	association constant
k_2	dissociation constant
K	equilibrium or affinity constant
Logit x	$\text{Log}_e \frac{x}{1-x}$
T	total radioactivity in each assay tube

1 Introduction

Competitive binding techniques such as radioimmunoassay (RIA) are widely used to measure an enormous variety of compounds in biological fluids. Current methods have arisen from the pioneering work of Yalow and Berson[1] in the U.S.A. and Ekins[2] in the U.K. Much of the early development was concerned with the analysis of protein hormones, and nearly a decade passed before attention focussed also on small molecules such as steroids and drugs. The potential of immunoassay methods for drug monitoring in clinical and forensic laboratories and in addict treatment programmes resulted in the commercial production of immunoassays for various therapeutic and abused drugs, making the technique available to laboratories lacking the facilities to raise their own antisera and synthesise labelled compounds. However, commercial assays are not only expensive but are restricted in range, and so it is advantageous for a forensic laboratory to have the capability to devise "in-house" immunoassays suited to its particular requirements.

This chapter describes the theory and practice of RIA in forensic drug analysis. Much of the theory and some of the practice are applicable to immunoassays in which non-isotopic labels are used, but such assays are not described in detail since, to date, the versatility and sensitivity of RIA have made it the immunoassay technique of choice in forensic toxicology.

The particular advantages of RIA are its sensitivity and the fact that samples such as haemolysed blood can be assayed with little or no prior preparation. The worst samples often arise from the more serious cases, their size and type being dictated by the circumstances of the case rather than the choice of the analyst. The analyst must therefore be able to extract the maximum amount of information from, for instance, small, semi-solid, decomposing samples recovered from a long-dead body. It is in such cases that RIA is particularly valuable since a comprehensive RIA screen for drugs of abuse can be carried out quickly on a minimum amount of material, leaving the bulk of the samples available for further analysis and the confirmation of positive RIA results.

A further advantage of RIA is its ready applicability to drugs that are not easily detected by routine chromatographic methods. Examples include digoxin and LSD which are normally found in very low ng/ml concentrations in blood, benzoylecgonine which is difficult to extract from body fluids due to its polarity, and insulin which, because it is a protein hormone rather than a low molecular weight compound, is virtually undetectable in body fluids except by immunoassay methods.

The literature pertaining to RIA and other immunoassays is vast but various reviews[3–14] and books[15–26] offer a useful introduction while computerised literature searching provides easy access to the more specialised publications. The novice should not be deterred by the sheer volume of the literature since the basic principles are easily grasped and, if commercial kits are used initially, the necessary practical experience is readily acquired.

2 Theory

2.1 Basic Principles

In this section, the theory of competitive binding assays is outlined in sufficient detail for the basic principles to be understood.

Consider the addition of a sample or standard containing a variable concentration of unlabelled antigen, Ag, to a mixture of a fixed amount of labelled antigen, Ag*, and a fixed amount of antibody, Ab, that is sufficient to bind only part of the Ag* present. Antigen-antibody binding is a reversible bimolecular reaction that is governed by the Law of Mass Action and so an equilibrium results. This can be represented by

$$[\mathrm{Ag} + \mathrm{Ag}^*] + [\mathrm{Ab}] \underset{k_2}{\overset{k_1}{\rightleftharpoons}} [\mathrm{AgAb} + \mathrm{Ag}^*\mathrm{Ab}] \tag{1}$$

where k_1 and k_2 are the association and dissociation constants respectively and the square brackets indicate molar concentrations in the mixture according to the usual convention.

The equilibrium or affinity constant of the reaction, K, is given by

$$K = \frac{k_1}{k_2} = \frac{[\mathrm{AgAb} + \mathrm{Ag}^*\mathrm{Ab}]}{[\mathrm{Ag} + \mathrm{Ag}^*]\,[\mathrm{Ab}]} \tag{2}$$

The units of K are thus litres/mole.

At equilibrium, unbound antigen, Ag + Ag*, is known as the "free fraction" while antibody-bound antigen, AgAb + Ab, is known as the "bound fraction".

The unlabelled antigen, Ag, is the only variable in the system and so changes in its concentration affect the position of the equilibrium, an increase in [Ag] moving the equilibrium to the right and *vice versa*. The purpose of the labelled antigen, Ag*, is to enable the equilibrium position to be determined by measuring the distribution of the label between the bound and free fractions, and this is the basis of a competitive binding assay.

In RIA, the bound and free fractions have to be separated (Sect. 6) before the Ag* distribution can be measured, but systems requiring no separation step have been devised using non-isotopic Ag*. Assays are classified as "heterogeneous" or "homogeneous" depending on whether or not they require a separation step.

Quantitation in RIA is effected by assaying a series of standards along with the unknowns. When equilibrium is reached, the separation procedure is carried out and the radioactivity in the bound or free fractions (sometimes in both) is measured. "Totals", which are tubes containing only Ag*, are included in the assay so that the proportion of Ag* bound by Ab in the other assay tubes can be calculated. The next step is to construct a calibration curve which is often referred to as a standard curve or a dose-response curve. Various ways of plotting calibration curves have been compared[27]. The standard concentrations are plotted linearly or logarithmically on one axis, usually the abscissa, while the ordinate is a function of the measured radioactivity, e.g. counts per minute (cpm) in the bound or free fraction, $\%B^*/T$, $\%B^*/B_0^*$, B^*/F^*, logit $B^*/B_0^* \left(= \log_e \frac{B^*/B_0^*}{1 - B^*/B_0^*} \right)$ etc. where B^*, F^* and T are the measured activities in the bound fraction, free fraction and total ($= B^* + F^*$) respectively, and B_0^* is the measured activity in the bound fraction of the zero tube of the calibration curve, i.e. the activity bound by the antibodies in the absence of unlabelled antigen, Ag. The concentrations of the unknowns are then extrapolated from the resulting curve.

Nowadays assay data are generally processed by a microcomputer linked to the radioactivity counter and computer programmes of varying sophistication are available. Some programmes employ coordinates that give an approximately linear calibration curve[28] while others use curve-fitting procedures such as the spline function[29]. The choice of programme is not critical provided there are sufficient standards to define the shape of the calibration curve and as long as the errors associated with the different regions of the curve are known. Guidelines for immunoassay data processing have been the subject of a special report[30].

2.2 The Equilibrium Constant

The equilibrium or affinity constant, K, is a fundamental parameter of immunoassay theory. Its magnitude may be determined from a Scatchard plot[31] whose equation is derived in immunoassay terms as follows.

Rewriting Eq. (1) in its simplest form with no distinction between Ag and Ag*

$$Ag + Ab \rightleftharpoons AgAb \tag{3}$$

and

$$K = \frac{[AgAb]}{[Ag]\,[Ab]} \tag{4}$$

Substituting B for [AgAb], which is the concentration of antibody bound antigen, and taking $[Ag_T]$ and $[Ab_T]$ as the total concentrations of Ab and Ag in the system, Eq. (4) becomes

$$K = \frac{B}{[Ag_T - B]\,[Ab_T - B]} \tag{5}$$

Substituting F for $[Ag_T - B]$, which is the concentration of unbound or free Ag, and multiplying both sides by $[Ab_T - B]$ gives

$$K\,[Ab_T - B] = \frac{B}{F} \tag{6}$$

and so

$$\frac{B}{F} = K\,[Ab_T] - KB \tag{7}$$

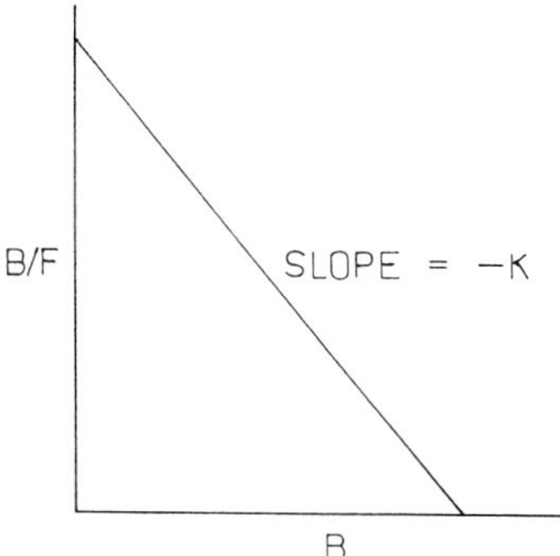

Fig. 1. Idealised Scatchard plot

Both K and Ab_T are constants and so a plot of $\frac{B}{F}$ against B, with both B and F expressed in moles/litre, gives a straight line of slope −K. As mentioned earlier, the units of K are litres/mole. Fig. 1 shows an idealised Scatchard plot which expresses in a linear form the first order Mass Action Law governing the interaction between small molecules and proteins. It was applied to RIA data by Berson and Yalow[32].

As well as giving the value of K, a Scatchard plot can also be used to derive the total molar concentration of antibody binding sites, $[Ab_T]$, in the system. When $\frac{B}{F}$ is zero, Eq. (7) reduces to

$$B = [Ab_T] \tag{8}$$

and so $[Ab_T]$ is given by the intercept of the Scatchard plot with the B axis. The concept of a molar concentration of binding sites need not be confusing if it is born in mind that one mole of a binding site is the amount required to bind one mole of antigen with 100% saturation. Conversion to w/v units is possible since an IgG antibody molecule has two identical binding sites and a molecular weight of 140,000, i.e. one mole of an IgG binding site weighs 70,000 g.

The experimental data for a Scatchard plot are obtained by a procedure similar to that for constructing an RIA calibration curve but there is one important difference, namely that the amount of Ag* is proportional to the amount of Ag. For instance, if 20 ng/ml Ag contains sufficient Ag* to give 50,000 cpm, then 10 ng/ml Ag should contain sufficient Ag* to give 25,000 cpm. The ratio $\frac{B}{F}$ is thus obtained directly by measuring the activity in the bound and free fractions. If the Ag* concentrations are significant compared to those of Ag, then they should be included in the calculations though, with Ag* of high specific activity, they may perhaps be ignored.

In practice, a Scatchard plot is usually a curve rather than a straight line. This is because the theoretical model includes various assumptions, some or all of which may not be valid. These assumptions are that all Ag molecules are identical and possess only one epitope, that Ab and Ag* behave in an identical manner, that all Ab molecules are identical with one binding site apiece, that Ag and Ag reach equilibrium according to the first order Mass Action Law with second order chemical kinetics, and that B and F

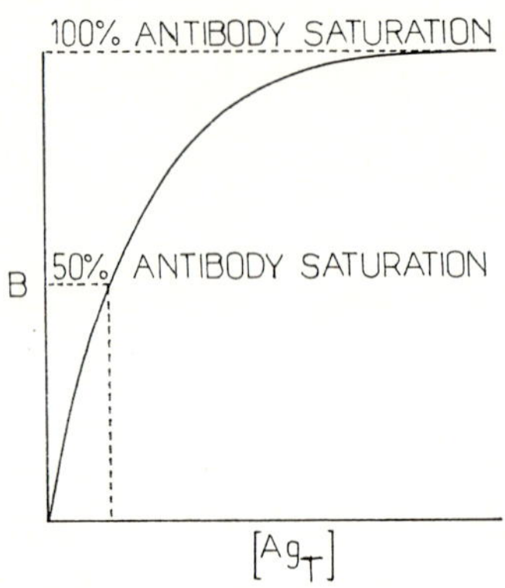

Fig. 2. Antigen-antibody saturation curve

can be separated and measured with 100% efficiency and without disturbing the equilibrium.

Provided the non-linearity of a Scatchard plot obtained in practice is not too extreme, a straight line can be fitted by eye or by the least squares method and estimates of K and $[Ab_T]$ can be derived.

An alternative method of estimating K has been proposed by Odell et al.[33]. Aliquots of Ab are equilibrated with varying amounts of Ag each containing the same proportion of Ag* and, after separating and measuring B and F, B is plotted against $[Ag_T]$ to give a saturation curve (Fig. 2). At 50% saturation of Ab, F is equal to 1/K and so K can be calculated. The relationship is verified as follows.

At 50% saturation of Ab, $[Ab_T]$ in Eq. (5) is equal to 2B and so

$$K = \frac{B}{[Ag_T - B]B} \tag{9}$$

Therefore, since $[Ag_T - B] = F$, which is the concentration of free Ag,

$$K = \frac{1}{F} \tag{10}$$

A polyclonal antiserum contains antibodies of varying avidity and so the K value obtained from a Scatchard plot or by the saturation method is an estimate of the average properties of the antiserum. The deviation from linearity of a Scatchard plot is largely due to this antiserum heterogeneity, the steeper portion of the curve being due to the high K value antibodies in the mixture.

Despite the divergence between the theoretical model and the practical reality of antigen-antibody binding, K values are useful for comparing antisera and assessing their likely value in an immunoassay. For example, Fig. 3 shows the Scatchard plots for two antisera, I and II. The slopes of the lines and the intercepts on the B axis show that antiserum I has a higher K value but is present in a lower concentration than antiserum II.

As will be shown later, the K value of an antiserum can also be used to estimate the potential sensitivity of an assay in which the antiserum is used.

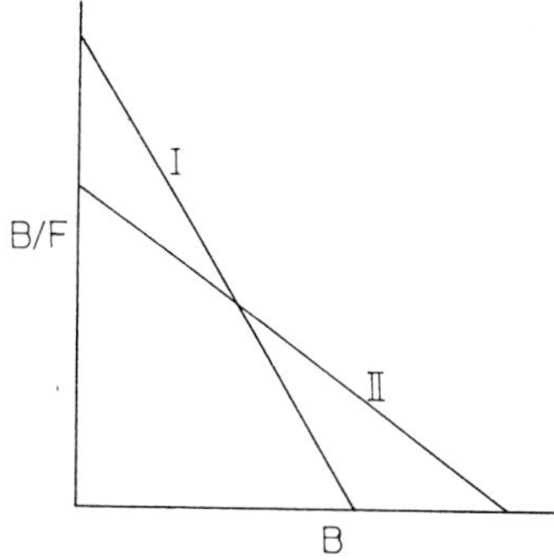

Fig. 3. Scatchard plots of two antisera, I and II

Typical K values of good antisera are of the order of 10^9–10^{11} litres/mole. K values can vary with temperature[34] but the change is likely to be minimal with antibodies to small molecules.

A word of caution is necessary. The measurement of the K values of drug antisera is likely to be more valid using ^{3}H-labelled rather than ^{125}I-labelled Ag*. This is because ^{125}I-labelled drug analogues are unlikely to be bound with the same avidity as the unlabelled drugs. Antisera for ^{125}I-labelled drug assays should be evaluated by their performance in the assay and not judged solely by their estimated K values.

2.3 Theoretical Immunoassay Model and Optimisation of Assay Conditions

A theoretical model of an immunoassay can be derived from basic principles and used to predict the effect of changes in the system. The model is based on the assumptions given in the previous section and is thus an approximation of the real situation. It is derived as follows.

From Eq. (4)

$$K[Ag][Ab] = [AgAb] \tag{11}$$

and, by definition,

$$\frac{B}{F} = \frac{[AgAb]}{[Ag]} \tag{12}$$

If, as before, $[Ag_T]$ and $[Ab_T]$ are the total concentrations of Ag and Ab in the systems, then

$$[Ag] = [Ag_T - AgAb] \tag{13}$$

and

$$[Ab] = [Ab_T - AgAb] \tag{14}$$

and so

$$\frac{B}{F} = \frac{[AgAb]}{[Ag_T - AgAb]} \tag{15}$$

From which

$$[AgAb] = \frac{B/F\,[Ag_T]}{(B/F + 1)} \tag{16}$$

Eqs. (11), (13), (14) and (16) are combined to give

$$K\left([Ag_T] - \frac{B/F\,[Ag_T]}{(B/F + 1)}\right)\left([Ab_T] - \frac{B/F\,[Ag_T]}{(B/F + 1)}\right) = \frac{B/F\,[Ag_T]}{(B/F + 1)} \tag{17}$$

Dividing by $[Ag_T]$ gives

$$K\left(1 - \frac{B/F}{(B/F + 1)}\right)\left([Ab_T] - \frac{B/F\,[Ag_T]}{(B/F + 1)}\right) = \frac{B/F}{(B/F + 1)} \tag{18}$$

and multiplying by (B/F + 1) gives

$$K(B/F + 1 - B/F)\left([Ab_T] - \frac{B/F\,[Ag_T]}{(B/F + 1)}\right) = B/F \tag{19}$$

from which

$$K\left([Ab_T] - \frac{B/F\,[Ag_T]}{(B/F + 1)}\right) = B/F \tag{20}$$

Multiplying again by (B/F + 1) gives

$$K(B/F + 1)[Ab_T] - K.B/F\,[Ag_T] = B/F\,(B/F + 1) \tag{21}$$

Rearranging the terms gives

$$B/F\,(B/F + 1) + K.B/F\,[Ag_T] - K(B/F + 1)[Ab_T] = 0 \tag{22}$$

from which

$$(B/F)^2 + B/F\,(1 + K[Ag_T] - K[Ab_T]) - K[Ab_T] = 0 \tag{23}$$

This is the Ekins equation[35], a simple quadratic that defines a hyperbola with B/F and $[Ag_T]$ as the variables. The concentrations (expressed in moles/litre if K is in litres/mole) are those in the assay mixture at equilibrium.

The labelled and unlabelled antigens, Ag and Ag*, must be considered separately if the model is to be used to compare assay conditions and so, with [Ag] + [Ag*] substituted for $[Ag_T]$, Eq. (23) becomes

$$(B/F)^2 + B/F(1 + K[Ag] + K[Ag^*] - K[Ab_T]) - K[Ab_T] = 0 \tag{24}$$

In designing an assay using Eq. (24), changes in the antibody and labelled antigen concentrations, $[Ab_T]$ and [Ag*], can be evaluated by varying the concentration of unlabelled antigen, [Ag], and calculating the effect on B/F, i.e. calibration curves are constructed for various combinations of $[Ab_T]$ and [Ag*]. Out of the possible permutations, two alternatives have been proposed for assays requiring high sensitivity, with a third alternative for assays in which low rather than high sensitivity is desired. The three alternatives are as follows.

2.3.1 Yalow and Berson's Conditions

Yalow and Berson[36] developed a mathematical RIA model somewhat different from that given above. They concluded that an antiserum concentration sufficient to bind 33% of a vanishingly small amount of Ag* in the absence of Ag would give the steepest calibration curve and hence the most sensitive assay. High specific activity Ag* is required so that the tiny amount used in the assay gives a statistically significant number of counts in a reasonable time.

Under these conditions, B/F = 0.5 and $[Ag] + [Ag^*] \cong 0$, and Eq. (24) becomes

$$0.5^2 + 0.5(1 - K[Ab_T]) - K[Ab_T] = 0 \tag{25}$$

and so

$$[Ab_T] = \mathbf{1}/2K \tag{26}$$

Thus, in any assay conforming to Yalow and Berson's conditions, the antibody binding site concentration, $[Ab_T]$, is inevitably 1/2K.

2.3.2 Ekins' Conditions

Ekins et al.[35, 37, 38] considered that the theoretical basis of Yalow and Berson's assay model was fallacious and developed an alternative which took into account the experimental error in determining B/F. They concluded that the conditions for maximum sensitvity were $[Ag^*] = 4/K$ and $[Ab_T] = 3/K$. Substituting these values into Eq. (24) with Ag = 0 gives

$$(B/F)^2 + B/F(1 + 4K/K - 3K/K) - 3K/K = 0 \tag{27}$$

from which

$$(B/F)^2 + 2B/F - 3 = 0 \tag{28}$$

from which

$$B/F = 1 \text{ (or } -3) \tag{29}$$

Thus, under Ekins' conditions, 50% of the Ag* is bound in the absence of any Ag regardless of the value of K. The specific activity of Ag* need not be particularly high. For instance, with $K = 10^{10}$ litres/mole and a labelled drug antigen, Ag*, of molecular weight 400, 40 pg of Ag* would be required per assay tube assuming an incubation volume of 250 μl, i.e. 250 μl of an Ag* concentration of 4/K. With one atom of ^{125}I per Ag* molecule (= 100% specific activity) and a counting efficiency of 70%, the 40 pg of Ag* would give approximately 3.4×10^5 cpm. However, 10^4 cpm per assay tube is an acceptable counting rate and would, in this example, be given by 40 pg of Ag* of about 3% specific activity.

In order to set up an assay using Ekins' conditions, the value of K must be known. A 4/K concentration of Ag* in the assay mixture is used and the antiserum dilution that binds 50% of this in the absence of Ag must have the required binding site concentration of 3/K.

It should be noted that the common practice of using an antiserum dilution that binds 50% of a convenient but arbitrary amount of Ag*, e.g. the amount giving 10^4 cpm, does not satisfy Ekins' conditions unless, by chance, the Ag* concentration in the assay mixture is 4/Kmoles/litre.

2.3.3 High Antiserum Concentration

An insensitive assay is required for measuring analytes present in relatively high concentrations in order to avoid the inconvenience and error involved in analysing very small samples or diluting the samples before analysis. Such an assay requires a high concentration of antibody binding sites of the order of 50/K and a 20–30% excess of Ag*, i.e. an Ag* concentration of about 60/K to 65/K.

With $[Ab_T] = 50/K$, $[Ag^*] = 1.3\,Ab_T$ and $[Ag] = 0$,

Eq. (24) becomes

$$(B/F)^2 + B/F(1 + 65 - 50) - 50 = 0 \tag{30}$$

from which

$$B/F = 2.675 \tag{31}$$

Thus 72.8% of the Ag* is bound in the absence of Ag. The conditions of this type of assay are not particularly critical and an Ag* of low specific activity may be used.

2.4 Precision, Sensitivity and Accuracy

2.4.1 Precision

Ekins[27, 38–40] has considered in some detail the concepts of sensitivity and precision as applied to RIA. The term precision is taken to mean reproducibility. It depends on the experimental error, the counting error and the slope of the calibration curve. These parameters vary over the range of the calibration curve and so, therefore, does the precision, a property known as "heteroscedasticity". The heteroscedasticity of an

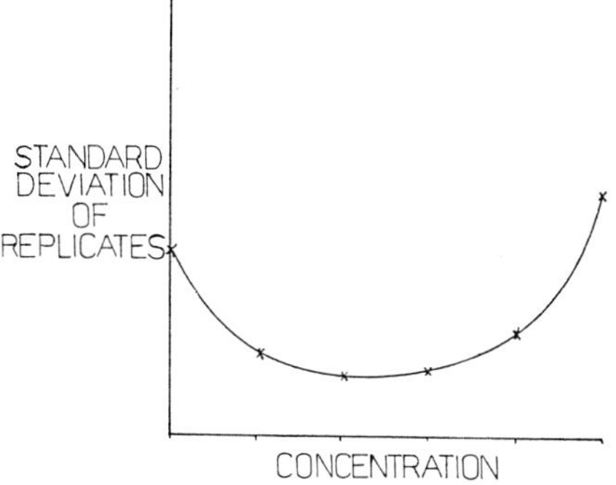

Fig. 4. Precision profile

immunoassay is best illustrated by running a number of replicates for each point on the calibration curve and plotting the standard deviations of the replicates against the concentrations of the standard solutions. Such a diagram is known as a precision profile (Fig. 4). Its practical value is that it delineates the region of the calibration curve where the the results will have the highest precision. Compared with quantitative chromatographic techniques for instance, RIAs are relatively imprecise with standard deviations of around 5–15%.

2.4.2 Sensitivity

Sensitivity is taken to mean the detection limit of an assay, i.e. the smallest amount of analyte that can be distinguished from zero with a high degree of confidence. The maximum possible sensitivity for an optimised RIA (assuming a negligible counting error) is E/K, where E is the experimental error[38]. For example[38], in a testosterone assay in which K was 3×10^8 litres/mole, which is equivalent to about 1 ml/ng, and E was 2%, E/K was 0.02/1 ng/ml, i.e. 20 pg/ml which was approximately the sensitivity found in practice when the assay was working well.

The concentration, E/K, refers to the whole assay mixture and not to the sample. If sensitivity is expressed as the analyte concentration in the sample or as an amount of analyte per assay tube, it is helpful to state the sample volume as well.

Sensitivity is inversely proportional to K and so it is limited by the quality of the antiserum as well as the magnitude of the experimental error. For instance, with a drug of molecular weight 300, an experimental error of 10% and two antisera with K values of 3×10^7 and 3×10^9 litres/mole, the values of E/K are 1 ng/ml and 10 pg/ml respectively. Calculating E/K is thus a useful indication whether or not an antiserum would be suitable for assaying a specified analyte concentration.

Sensitivity is sometimes expressed as the amount of analyte that inhibits the binding of Ag* by 10%, an arbitrary convention that has little practical value.

In practice, the sensitivity of an assay may be limited by the sample "background", making it impossible for the theoretical sensitivity to be attained. Endogenous compounds in the sample with structural similarities to the analyte may be bound by the antiserum, thus giving an erroneously high result. Alternatively, the labelled antigen may be bound by protein in the sample, and there is evidence[41–46] that some drug users have endogenous antibodies or binding proteins in their blood. This is equivalent to an increase in the antiserum concentration and can give rise to an erroneously high or low result depending on whether or not the method used to separate the bound and free fractions (Sect. 6) affects the endogenous protein.

The estimation of assay sensitivity thus becomes a practical rather than a theoretical procedure. One approach is to use analyte standards made up in the biological fluid to be analysed. Logically, this should cancel out interference in the assay by the sample matrix but, since biological fluids vary in composition and hence in their response in the assay, the problem may be compounded rather than eliminated. An alternative approach is to prepare the standards in the assay buffer, thus ensuring a stable calibration curve. A statistically significant number (> 30) of blank samples which are usually blood or urine are then assayed and the sensitivity or detection limit of the assay is taken as the mean response of the blank samples plus three standard deviations. This eliminates the risk of false positive results with virtually 100% certainty. There is of course a chance that some samples giving results below the detection limit are in fact positive but, with a sensitive assay, the analyte levels in such samples will be well below these required to produce a significant physiological response. Should it be necessary to demonstrate as far as possible the absence rather than the presence of a drug and an RIA result below but close to the detection limit indicates a possible low positive, then further analysis is necessary and methods of high sensitivity, e.g. gas chromatography-mass spectrometry with single ion monitoring, must be used.

The non-specific effect of endogenous sample constituents usually depends on weak binding forces and does not mimic the analyte response exactly. Dilution of the sample or competition with the stronger antigen-antibody binding forces reduces the relative contribution of the non-specific effect and so, in assays in which non-specific effects are most marked, estimation of the detection limit in terms of the response of blank samples may be invalid. In such assays the non-specific effect can usually be eliminated by extracting the analyte prior to analysis, but fortunately this is not often necessary.

2.4.3 Accuracy

The accuracy of an assay is a measure of how well a result agrees with the true value of the analyte concentration. It is easily checked by comparing the results with those of an alternative and unrelated method such as gas chromatography or high performance liquid chromatography. Analysing spiked samples is another way to check the accuracy of an assay with the proviso that spiked and "real" samples may differ, e.g. in protein binding of the analyte or in the erythrocyte/plasma analyte distribution, and may not be strictly comparable.

Accuracy is a less meaningful concept in an assay that responds to metabolites as well as to the parent drug. With such an assay, precision and sensitivity are more important, the essential criterion being the ability to distinguish between positive and blank samples.

3 Practical Design of an Assay

The theoretical model of an immunoassay is based on the assumptions given in Section 2.2. In practice, particularly with drug RIAs, some of the assumptions are such poor approximations to reality that the theory becomes no more than a rough guideline to the empirical design of an assay.

To illustrate the practicalities of assay design, let us assume that we have a radioiodinated drug antigen or radioligand, a polyclonal antiserum to which it binds and a separation method (Sect. 6) capable of separating the bound and free fractions. A low ionic strength buffer of approximately neutral pH is required for diluting the radioligand and antiserum, e.g. 0.1 M phosphate, pH 7.4, containing 0.1% sodium azide as a preservative. It is often necessary to include some protein in the buffer, e.g. 0.1% w/v bovine gamma globulin or bovine serum albumin, to act as a carrier in the separation stage and/or to reduce non-specific binding of the radioligand to the walls of the assay tubes. Also required are pipettes with disposable tips, assay tubes such as 1.5 ml polypropylene microcentrifuge tubes or 3 ml round-bottomed polystyrene tubes, suitable racks, a vortex mixer, a centrifuge and a gamma-counter.

Development of an assay includes running an antiserum dilution curve to determine the working dilution of the antiserum, constructing and optimising a calibration curve and then assessing the usefulness of the assay by estimating the precision, recovery and detection limit and investigating the cross-reactivities of metabolites and other compounds resembling the analyte. Some of the procedures vary depending on whether the assay is "specific" or "general", i.e. whether it is designed to detect either a single drug or a range of structurally-related drugs.

3.1 Antiserum Dilution Curve

The first step in setting up an antiserum dilution curve is to dilute a portion of radioligand in buffer so that a convenient aliquot for use in the assay (50–100 µl) gives about 10^4 cpm. The standard deviation of a radioactivity count is the square root of the count and so the coefficient of variation of 10^4 cpm is 1%. Aliquots of radioligand are then incubated in duplicate with 50–100 µl aliquots of various dilutions of the antiserum. Extra buffer (50–100 µl) is added to allow for the samples or standards that will be used in the assay. An incubation time of one hour at room temperature is normally sufficient for a drug assay. The bound and free fractions are then separated and, after counting, the antiserum dilution curve is plotted. A typical protocol follows and dilution curves are shown in Fig. 5.

Protocol for antiscrum dilution curve:

Duplicate sets of tubes containing:—

Total

50 µl radioligand (10^4 cpm)

No further treatment other than counting

Antiserum

50 µl buffer

50 µl radioligand (10^4 cpm)

50 µl antiserum. Dilution factors of 1, 2, 4, 8, 16 ... may be used.

Vortex mix
Incubate at room temperature 1 h
Add 300 μl saturated ammonium sulphate in distilled water
Vortex mix (thoroughly)
Centrifuge 2 min 10^4 g
Aspirate or decant supernatants
Count precipitates (bound fractions)

Non-specific binding (NSB)
As above using 50 μl buffer instead of antiserum.

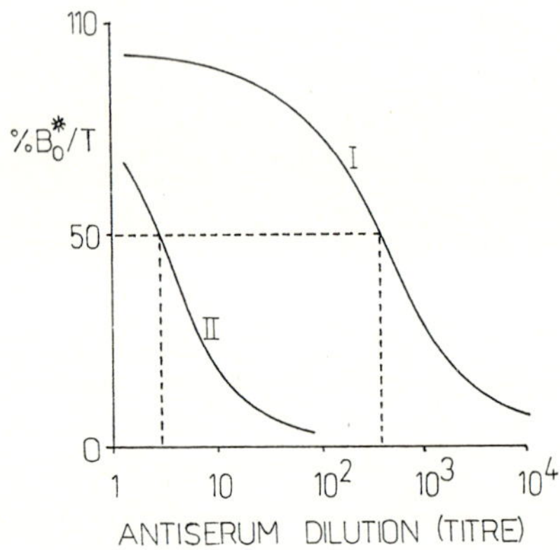

Fig. 5. Good (I) and poor (II) antiserum dilution curves

An antiserum dilution curve is usually sigmoid. Most of the radioligand should be bound by a low dilution (high concentration) of antiserum providing the label is reasonably pure and the antiserum sufficiently avid. With a high dilution (low concentration) of antiserum, the curve flattens out at a level approximating to the non-specific binding (NSB) which is a measure of the amount of label that adheres to the walls of the tubes or is trapped in the bound fraction but not actually bound by the antibodies. The NSB should not exceed 10–15% and should preferably be less than 5% B_0^*/T.

As a general rule, it is worth proceeding further if more than 80% of the radioligand can be bound by the antiserum with an NSB of less than 15%, and if 50% of the radioligand is bound by an antiserum dilution of 30–50 or greater. The dilution of antiserum required to bind 50% of the radioligand is referred to as the "titre" and is a useful parameter in evaluating or comparing antisera. A titre of less than 1:30 could be used but there is little point in doing the substantial work required to develop an assay unless the antiserum will last for several years or longer. For example, 20 ml of antiserum (approximately one bleed from a rabbit) with a titre of 1:10 is sufficient for 4000 assay tubes at 50 μl/tube. The antiserum will last for less than a year if it used at a rate of 100 tubes/week but, if the titre is $1{:}10^3$, it will last for more than 70 years.

The titres of drug antisera are apt to differ widely; those in use in the author's laboratory, for example, range from 1:30 to $1{:}10^4$.

An excessive NSB may sometimes be improved by changing the type or even the batch of the assay tube, increasing the protein concentration in the buffer, including detergent[47] or alcohol[48] in the assay mixture, extracting the analyte from the sample to eliminate endogenous binding proteins, or adopting an alternative separation procedure.

Little can be done to improve a poor antiserum dilution curve other than trying different bleeds of antiserum or alternative radioligands. It is wise to check that antigen-antibody equilibrium has been attained by increasing the incubation time. Changing the buffer or pH, lowering the incubation temperature or using different relative proportions of antiserum and radioligand may be help a little, while an alternative separation procedure occasionally effects a dramatic but unpredictable improvement.

3.2 Calibration Curve

If the antiserum dilution curve is satisfactory, the next step is to construct a calibration curve using the antiserum titre that binds 50% B_0^*/T.

The radioligand and antiserum are incubated with various concentrations of unlabelled drug in buffer. The protocol employed for the antiserum dilution curve is used and each drug standard is run in duplicate. Total, but not NSB, tubes are required. The results are plotted e.g. as shown in Fig. 6.

There are various possibilities. The curve may be too shallow (Fig. 6, Curve I) and thus of little practical value since the precision will be poor. Standards covering a wider concentration range will improve the curve but the assay will not be sensitive enough to detect therapeutic levels of the drug. Alternatively the curve may be too steep (Fig. 6, Curve III), flattening out at concentrations below the therapeutic range. Finally, the slope of the curve may extend over the therapeutic range and it is neither too shallow nor too steep (Fig. 5, Curve II).

Further development depends on whether the assay is "specific" or "general", i.e. whether it responds primarily to a single analyte or whether it cross-reacts with a range of structurally related analytes.

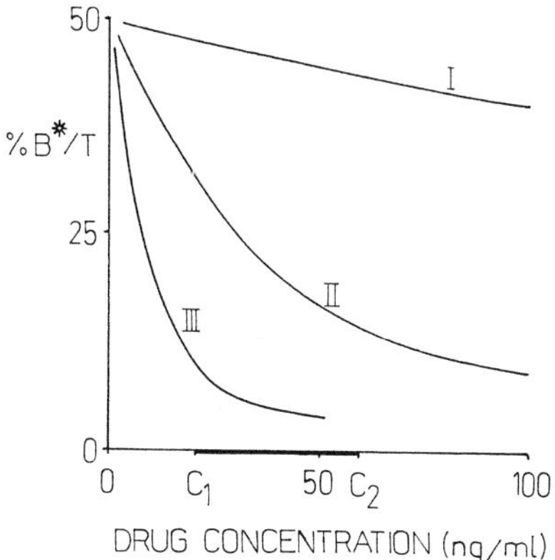

Fig. 6. RIA calibration curves. C_1—C_2 indicates the therapeutic range of the drug in blood

3.3 Specific Assay

An assay is specific if only the analyte is bound by the antiserum or if significant amounts of other cross-reacting compounds are never found in the samples to be analysed. Specific assays are widely used in clinical biochemistry and are useful too in forensic toxicology for compounds such as insulin, LSD and methadone that do not have commonly-encountered close structural analogues. Fewer variables are involved in the development of a specific rather than a general assay.

A shallow calibration curve indicates either that the antiserum binds the radioligand more avidly than it binds the unlabelled drug or that the K value of the antiserum is too low to permit the desired concentrations to be measured. Occasionally, the addition of extra protein, e.g. 50 μl serum/assay tube, can improve a marginally poor curve but, failing this, the only alternative is to try different antisera and radioligands, perhaps with heterologous bridge linkages (see Sect. 4.3).

A calibration curve that is too steep and therefore too sensitive is much less of a problem. Sensitivity is decreased and the standard concentration range is extended by using an antiserum titre that binds 70–80% of the radioligand in the absence of unlabelled drug.

A calibration curve that approximates to the desired shape may benefit from optimisation. The theoretical principles of assay optimisation (Sect. 2.3) will not hold for structurally dissimilar radioligands and analytes unless, by chance, they are bound with equal avidity by the antiserum, and so trial and error optimisation is necessary.

Decreasing the radioligand increases the sensitivity, but successive reduction has successively less effect and the counting time must be increased to avoid a loss of precision (Fig. 7).

Varying the antiserum titre has a marked effect on sensitivity (Fig. 8). Increasing the titre improves the sensitivity though the precision may suffer. As already mentioned, decreasing the titre lowers the sensitivity.

A radioligand of reasonably high specific activity is best for a sensitive assay while, for an assay of low sensitivity, a low specific activity is adequate and the radioligand may be diluted with unlabelled drug if necessary. Cost should be considered for it is cheaper to analyse diluted samples with a sensitive assay than to use antiserum and radioligand in the amounts required for a low sensitivity assay.

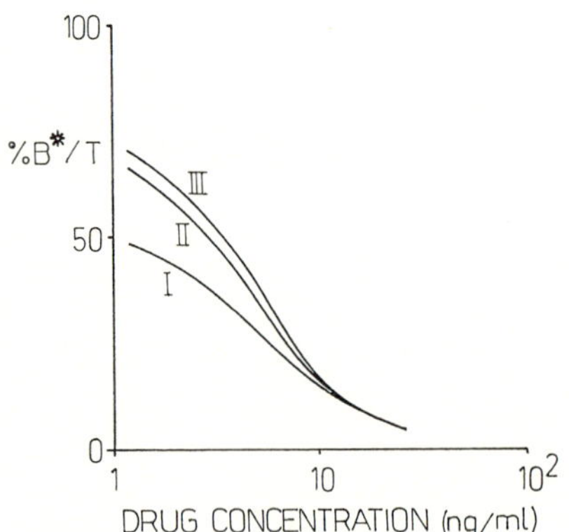

Fig. 7. Effect of reducing the radioligand. The total radioligand used to construct curves I, II and III was T, T/10 and T/100 respectively

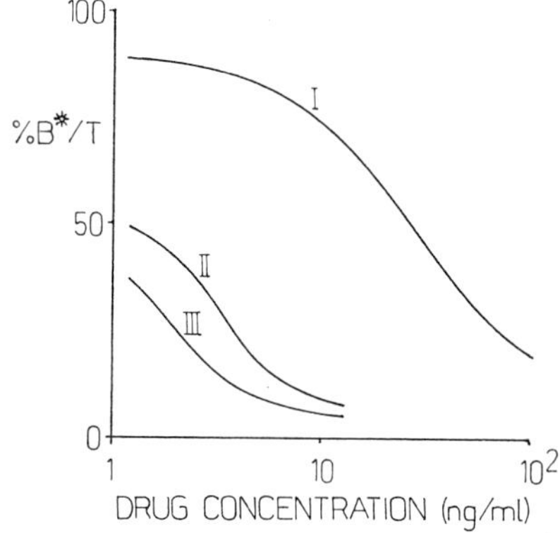

Fig. 8. Effect of increasing the antiserum titre. Curves I, II and III were constructed using low, intermediate and high antiserum titres respectively

A practical method of optimising the sensitivity is to run two antiserum dilution curves, one as already described and the second with the inclusion of a low concentration of unlabelled drug in the assay tubes[49]. The antiserum titre for the most sensitive assay is defined by the maximum vertical spacing between the curves (Fig. 9).

Increasing the amount of sample or standard in the assay mixture increases the sensitivity but the gain may be offset by an increased sample background.

Carrying out the separation before equilibrium is reached ("disequilibrium assay") can affect sensitivity, as can changing the order in which reagents are added to the assay tubes. Normally the standard or sample and the radioligand are added before the antiserum but, if the antiserum is allowed to react with unlabelled drug in the samples or standards before adding the radioligand ("late-addition assay")[50], the kinetics of the overall reaction are altered and, if the separation is carried out before equilibrium, the sensitivity may be increased. Whether or not a disequilibrium or late-addition assay has the desired effect must be tested empirically. Such assays suffer from the disadvantage that the timing is critical and so the precision is apt to be poor.

An acceptable calibration curve can often be established fairly rapidly by trial and error as described above. Further refinement may be possible by varying other parameters of the assay such as the incubation time and the pH.

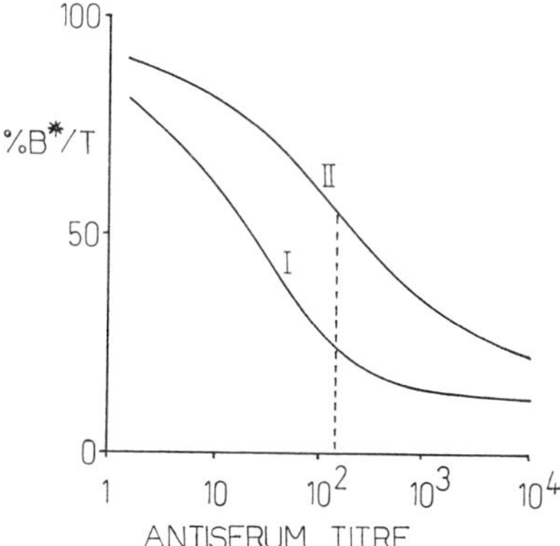

Fig. 9. Antiserum dilution curves with (I) and without (II) added unlabelled drug. The antiserum titre for maximum sensitivity is indicated by the dotted line

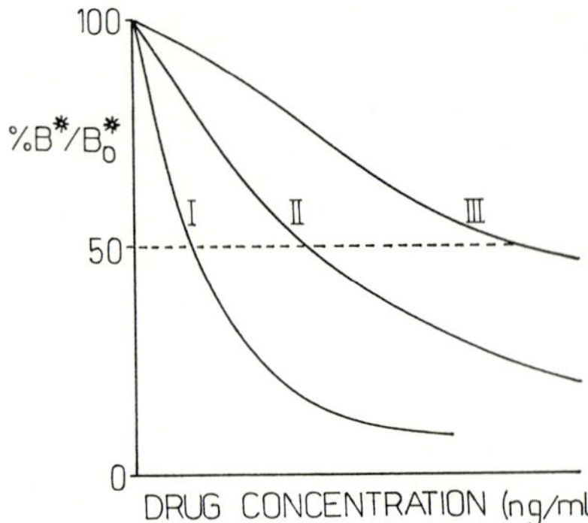

Fig. 10. Calibration curves of cross-reacting compounds

It is advisable to compare the calibration curves resulting from standards made up in buffer and in the sample matrix, e. g. blood, urine etc. The curves should be parallel but not necessarily coincident. Non-parallel curves indicate a matrix effect, perhaps due to endogenous binding proteins in the sample (Sect. 2.4.2), which may have to be eliminated by extracting the drug prior to analysis.

Having optimised the calibration curve, the specificity of the assay must be assessed. Metabolites and other structural analogues of the analyte are tested to see if they cross-react with the antiserum. Calibration curves of all such compounds are run and the results are plotted (Fig. 10). If % B^*/B_0^* is used for the ordinate, all the curves originate from the same point but, for a valid comparison, all curves should have the same initial binding, i.e. the same value of B_0^*/T.

Cross-reactivity data are usually expressed in terms of the amounts of the various analytes required to reduce the initial binding by 50%. Often the results are normalised with respect to a reference analyte, i.e. % cross-reactivity of analyte I = weight of reference analyte required to reduce initial binding by 50% × 100/weight of analyte I required to reduce initial binding by 50%. The curves are not parallel and may even cross, so the data are no more than estimates of the relative cross-reactivities of the analytes.

Absolute specificity in a drug immunoassay is highly unlikely since metabolites or structurally related drugs will probably cross-react, albeit to a small extent. For practical purposes, however, an assay can be considered specific if metabolites etc. are encounted in such low amounts or else cross-react so poorly that they have little effect on the measurement of the analyte of interest.

Specificity or lack of it is largely a property of the antiserum and so is not amenable to modification though, if more than one antiserum and/or radioligand is available, the optimum combination may be chosen. This procedure has been termed "chemical tuning"[51], though the degree of control over the process is somewhat lacking. A less serendipitous way of improving specificity is purification of the analyte prior to analysis. Extraction methods or chromotography may be used, but an elaborate purification procedure may detract from the speed and simplicity of RIA to the point where an alternative method of analysis becomes preferable or more cost-effective.

The detection limit, precision and accuracy of the assay must now be investigated. The detection limit is determined as detailed in Sect. 2.4.2 by averaging the response of blank samples and adding three standard deviations. Intra- and inter-assay precision is estimated by replicate analysis of spiked samples, and accuracy is checked against a reference method.

Finally, it is advisable to run the assay in tandem with an alternative method when undertaking case-work until the operator is satisfied that the results are valid.

3.4 General Assay

The concept of a general RIA is exclusive to forensic toxicology. Ideally, such an assay responds to all the drugs of a single category, e.g. barbiturates, benzodiazepines, opiates etc. Case samples can therefore be screened and positives identified rapidly and with minimum cost and effort. Further work is required to identify and quantitate the drugs found in positive samples, but the confirmation of positive results is already an established practice in forensic toxicology. In most instances, negative RIA results are unequivocal and require no further work.

The development of a general assay resembles that of a specific assay, but the aim is to find the radioligand-antiserum combination with the highest sensitivity over the broadest spectrum of cross-reactivity.

The antiserum with the highest titre is identified by running dilution curves. Using this antiserum, calibration curves are constructed for all drugs in the category under consideration. Three or four drugs spanning the whole cross-reactivity range can then be used to check other antisera with reasonable titres. The antiserum giving the best sensitivity over the cross-reactivity range is selected and the assay sensitivity is optimised, paying particular attention to the drugs with the lowest cross-reactivities.

The process is laborious, particularly when several antisera and radioligands have to be tested, but the long-term advantages of a good general assay are substantial.

Calibration curves of all the drugs are run on the optimised assay and the detection limit is determined by calculating the $\% B^*/B_0^*$ value of the mean blank sample response plus three standard deviations. This figure enables detection limits in ng/ml for all the drugs to be extrapolated from the calibration curves.

The practical value of the assay is assessed by comparing the detection limits with the therapeutic levels of the drugs. As an example, Table 1 summarises published data[52] for a tricyclic depressant RIA.

In practice, one drug is used as the assay standard and results are given as apparent concentrations of this drug. The drug chosen as the standard may be the most frequently encountered of its type or it may be the one with the highest cross-reactivity. The precision of the assay is estimated using samples spiked with this drug while recoveries should be checked for a representative selection of the drugs.

Accuracy is a somewhat meaningless concept when applied to a general RIA though precision is important, particularly the distinction between blank and positive samples.

Variations in structure make it unlikely that all drugs of one class will cross-react well in a general assay. Inevitably, therapeutic levels of certain drugs will not be detectable but, with reasonable luck, these will be drugs that are rarely encountered. However, the limitations of a general assay must not be forgotten in practice.

The cross-reactivity spectrum of a general assay can be improved by mixing antisera[55] but, to date, this promising field has been little explored.

An RIA that detects a single drug plus its metabolites is a type of general assay. Such an assay would have limited application in therapeutic drug monitoring but, in forensic

Table 1. Detection limits and therapeutic ranges of tricyclic antidepressants and related compounds[52]

	Detection Limit, ng/ml		Therapeutic Range
	Blood	Urine	in Blood, ng/ml[53, 54]
Amitriptyline	41	28	35 – 202 16 – 500
Butriptyline	31	22	60 – 280
Carbamazepine	6500	22000	3500 – 9400
Clomipramine	45	51	20 – 143 16 – 282
Cyproheptadine	17.5	34	n.k.
Desipramine	15	12.5	11 – 110 8 – 280
Dibenzepin	25	27	180
Dothiepin	36	25.5	17 – 64 25 – 420
Doxepin	25	21	28 – 171 5 – 115
Imipramine	23	17	9 – 126
Iprindole	52.5	84	n.k.
Maprotiline	20	42	26 – 196 45 – 1558
Mianserin	500	1500	30 – 120
Nortriptyline	36	23	46 – 253 10 – 275
Noxiptyline	19	29	n.k.
Opipramol	29	17.5	n.k.
Protriptyline	50	21	95 – 288 10 – 376
Trimipramine	60	47	310,600

n.k. = not known

toxicology, the ability to detect metabolites effectively increases the sensitivity of the assay and enables drug use to be detected when only metabolites remain in the sample.

3.5 Multiple Drug Assay

The detection of two or more chemically unrelated drugs with a single RIA is a logical alternative to using individual specific or general assays. Such assays have been developed by mixing antisera[56] and by immunizing animals with up to four different immunogens[57], and are best suited for large-scale drug screening where only a few positive results are expected.

4 Radioligands

The function of the radioligand in RIA is to permit measurement of the changes in the antigen-antibody equilibrium that occur as the concentration of unlabelled drug in the system is varied. It is not necessary for the radioligand and unlabelled drug to have exactly the same immunoreactivity, i.e. to be bound with equal avidity by the antiserum, but a radioligand that is strongly or weakly bound compared with the unlabelled drug can have an adverse effect on the sensitivity and precision of the assay.

4.1 β-Emitting Isotopes

The most commonly used β-emitter in RIA is ^{3}H, while ^{14}C is employed occasionally and ^{35}S (half-life 87.4 days) has been used to label the sulphur-containing drug, propylthiouracil[58]. The half-life of ^{3}H (12.26 years) is much shorter than that of ^{14}C (5730 years) and so ^{3}H-labelled compounds have considerably higher specific activities than the corresponding ^{14}C-labelled analogues. For example, at 100% isotopic abundance, ^{3}H has a specific activity of 1.07 TBq/milligramatom while that of ^{14}C is only 2.31 GBq/milligramatom. This several hundredfold difference means that the very small amounts of a ^{3}H-labelled radioligand required in RIA are easily quantitated by liquid scintillation counting, although the counting efficiency for ^{3}H (approximately 30–60%) is lower than that for ^{14}C (approximately 70–90%). While ^{14}C-labelled radioligands have been used in RIA, e.g.[59], the detection limits of such assays are generally poor due to the amount of radioligand required to accumulate a significant number of counts in a reasonable time. Changing to a ^{3}H-labelled radioligand effects a considerable improvement[60].

4.1.1 β-Labelled Drugs

Few laboratories have the facilities to synthesise and purify β-labelled drugs. Fortunately a number of ^{3}H-labelled drugs are routinely available from commercial organisations who also offer custom synthesis or a ^{3}H-labelling service for non-routine compounds. While the cost of a custom synthesis reflects the complexity of the procedure, the cost of ^{3}H-labelling by, for instance, isotope exchange reactions is not unreasonable. The specific activities of the products can vary widely depending on the efficiency of the labelling method and the number of ^{3}H atoms incorporated into each molecule. An alternative procedure that can be carried out "in-house" is to conjugate the material of interest to a commercially-available ^{3}H-labelled compound, e.g.[61].

4.1.2 Stability and Storage of β-Labelled Drugs

Isotopically labelled compounds are subject to various modes of decomposition which can be minimised by a suitable choice of storage conditions[62, 63]. Natural isotopic decay can lead to further decomposition if the α, β or γ emissions either interact directly with the molecules of the labelled compound or via radiolytically produced excited species or free radicals. Decomposition can also occur if the compounds are inherently unstable or are stored under conditions permitting hydrolysis, oxidation etc. to occur. The correct choice of storage conditions for labelled compounds is therefore important.

For drugs, storage of dilute solutions in an organic solvent such as ethanol at low temperature is generally effective.

4.1.3 Liquid Scintillation Counting

β-emitters such as ^3H are measured by liquid scintillation counting. This is because the β-particles, which are electrons, are rapidly attenuated by water or other solvents, glass and plastic and so cannot be detected directly in solution or in a glass or plastic container.

In liquid scintillation counting, the energy of a β-particle is absorbed by a liquid scintillant and then re-emitted as a flash of light which is registered by a linked pair of photomultiplier tubes. The liquid scintillant or scintillation "cocktail" consists of an aromatic solvent or mixture of solvents containing one or more highly-conjugated fluorescent aromatic solutes. Different formulations can accommodate a variety of sample types in solution, in suspension or, if a detergent is included, in emulsion. Decay of a ^3H atom in the sample produces a β-particle whose energy is transferred largely to the solvent molecules which form the bulk of the mixture. Energy is then transferred from the excited solvent molecules to the solute molecules which then fluoresce, re-emitting the energy as a cluster of photons or "scintillation" which is detected by the photomultiplier tubes. The wavelength range of the fluorescence spectrum of the solutes in the liquid scintillant should match the spectral response of the photomultiplier tubes (generally in the near UV/visible region, about 300–500 nm), and the concentration of the solutes (a few g/l) should be sufficient to ensure a virtually quantitative transfer of energy from the solvent to the solute molecules.

β-particles from ^3H or radiations from other isotopes have various energies up to a maximum that is characteristic of the isotope, and a plot of the number of radiations emitted in unit time against the energies of the emissions in electron volts gives the energy spectrum of the isotope (Fig. 11a).

In liquid scintillation counting, the number of photons in the scintillation produced by a β-particle depends on the energy transferred to the liquid scintillant by the particle. Each photomultiplier tube amplifies the scintillation and produces a pulse of electricity whose voltage or amplitude is proportional to the number of photons in the scintillation. The energy spectrum of the isotope is thus transformed into a "pulse

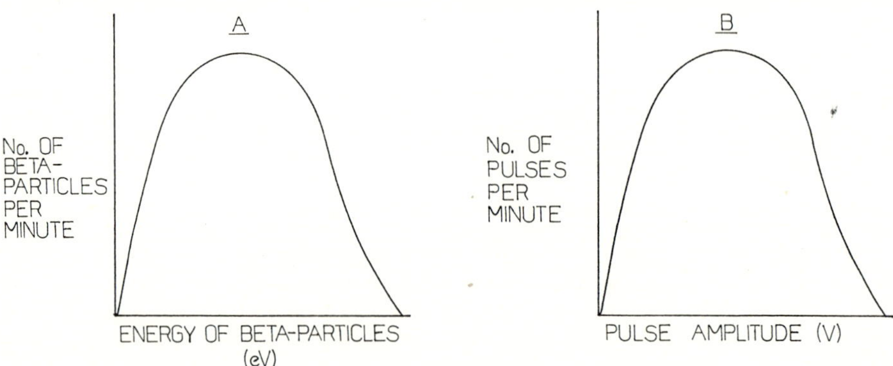

Fig. 11. Idealised energy (A) and pulse amplitude (B) spectra of a β-emitter

amplitude" spectrum in which the number of pulses in unit time is plotted against the pulse amplitude in volts (Fig. 11b).

The two photomultiplier tubes in a liquid scintillation counter are linked by coincidence circuitry so that only scintillations seen by both tubes simultaneously are accepted by the instrument. This minimises the "background" counts or "noise" due to cosmic rays etc. Coincident pulses are registered as "counts" by one or more "channels" each equipped with adjustable upper and lower gates or "thresholds" that define the acceptable pulse amplitude range, i.e. only pulses whose amplitudes lie between the voltages set by the upper and lower thresholds are counted in an individual channel. For counting a single isotope, one channel is used with the thresholds adjusted so that the count rate for the isotope is maximised while keeping the background to a reasonable minimum. Two isotopes can be counted simultaneously using two channels as long as the pulse amplitude spectra do not overlap completely (Fig. 12). In the figure, Channel 1 is set to count ^{3}H and some ^{14}C, while Channel 2 counts ^{14}C only. The results from Channel 1 are corrected using the results from Channel 2 to give an accurate measure of the ^{3}H present in the mixture.

Components of the sample may have a deleterious effect on the amplitude of the scintillation pulses, a phenomenon known as "quenching". There are three types of quenching; impurity (also known as chemical) quenching, colour quenching and photon quenching.

Impurity quenching occurs when some of the excited solvent molecules transfer their energy to non-fluorescent substances in the sample, the energy thus being dissipated by radiationless transfer. Mild impurity quenchers include organic chlorides, primary amines, sulphides and unsaturated aliphatic hydrocarbons, while severe quenching is caused by oxygen, organic bromides, iodides, nitro-compounds, primary mercaptans, acid chlorides and anhydrides, aldehydes, ketones, and secondary and tertiary amines.

Colour quenching occurs when scintillation photons are absorbed by coloured compounds in the sample. It can be reduced by diluting or bleaching the sample, or by including an additional solute in the liquid scintillant to shift the scintillation spectrum to a wavelength where the solution is more transparent.

Photon quenching arises if the sample combines incompletely with the liquid scintillant to give a heterogeneous mixture in which the emitted β-particles cannot interact fully with the solvent molecules.

Modern liquid scintillation counters not only have preset channels for counting different isotopes, but also incorporate quench correction facilities that may be largely

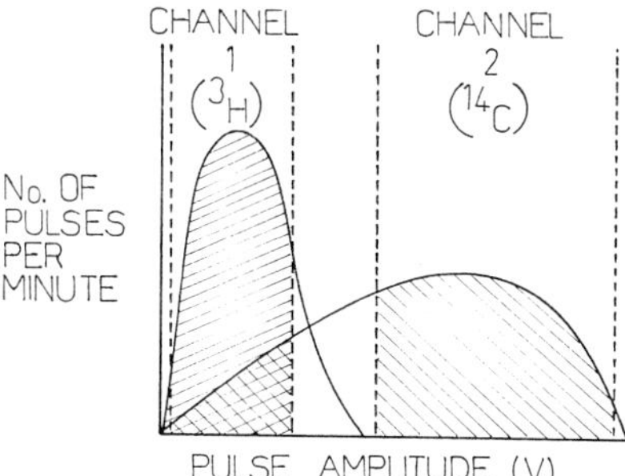

Fig. 12. Dual isotope counting

automatic. For a detailed discussion of quench correction, the reader is referred to the literature[19, 64–69].

While it is methodologically simpler to develop an assay with a ^{3}H rather than an ^{125}I label, the cost and complexity of liquid scintillation counting are disadvantageous in the long term. Haemolysed blood or other coloured samples must be extracted prior to analysis to avoid severe colour quenching and, if polyethylene glycol is used to separate the bound and free fractions (see Sect. 6), scintillants containing 1,4-dioxane, e.g.[70], rather than the cheaper toluene or toluene: Triton X-100 must be used to avoid incomplete incorporation of the sample. An alternative, which requires an extra step but minimises the overall cost, is to separate the bound and free fractions using polyethylene glycol and then extract the ^{3}H-labelled (and unlabelled) drug from an aliquot of the free fraction into an inexpensive toluene-based liquid scintillant (reagent-grade low-sulphur toluene containing 4 g/l 2,5-diphenyloxazole and 0.2 g/l 1,4-bis[2-(5-phenyloxazolyl)]-benzene). The free fraction (about 400 μl) is shaken for 10 minutes with 4 ml liquid scintillant in a 5 ml plastic liquid scintillation counting vial. The phases are then allowed to separate before counting. In this laboratory, the procedure has been used successfully in benzodiazepine[71, 72], cannabinoid[13, 14], methadone[73] and tricyclic antidepressant[13, 14] assays. Apart from the low cost, the counting efficiency (defined as the ratio of the observed counts per minute to the disintegrations per minute) is higher than that obtained with liquid scintillants that form emulsions or are miscible with aqueous samples. The counting time thus saved more than compensates for the extraction which takes 10 minutes on a laboratory shaker fitted with plastic foam racks for 80 tubes. A further time-saving measure is the use of a pneumatic dispenser/diluter to take up the 400 μl aliquots of the free fractions and dispense them into the counting vials along with 4 ml aliquots of liquid scintillant. Despite the immiscibility of the samples and scintillant, the coefficient of variation of the transfer is below 1%.

Liquid scintillant may also be used to separate the bound and free fractions of the assay mixture by extracting the free fraction directly from the equilibrium mixture[74, 75]. This enables the assay including counting to be carried out in single tubes with a concomitant time and cost saving. The experimental conditions must be carefully optimised to ensure reproducibility and to avoid disturbing the equilibrium.

The choice of instrument for liquid scintillation counting depends on the expected sample through-put and the money available, though automatic sample-changing is essential. Manufacturers' specifications are continually improving and so the prospective purchaser should critically compare available instruments that meet his requirements. The trend in recent years has been to use smaller volumes of sample and liquid scintillant, and so it is advisable to select an instrument that can accept different sizes of counting vials.

Most commercially available drug RIAs use the γ-emitter, ^{125}I, as the label and so liquid scintillation counting facilities are not absolutely necessary in a forensic laboratory. However, if development work is undertaken, a ^{3}H-labelled drug is very useful for testing antisera and, when a suitable antiserum has been selected, an assay can usually be set up fairly quickly using the ^{3}H-labelled drug to provide casework results while ^{125}I-labelled drugs are investigated. Testing antisera with a ^{125}I-labelled drug has the disadvantage that occasionally it may be bound by an antiserum to a much greater or lesser extent than the corresponding ^{3}H-labelled or unlabelled drug.

4.2 γ-Emitting Isotopes

The two γ-emitting isotopes that have been most widely employed in RIA are ^{131}I and ^{125}I, others such as ^{57}Co, ^{113m}In and ^{75}Se being used to only a limited extent[76–78]. Yalow and Berson used ^{131}I in their pioneering studies of insulin RIA[1] but ^{125}I is now the isotope of choice, since compared with ^{131}I, it has a more convenient half-life, it can be counted more efficiently, it emits less penetrating γ-radiation and it can be purchased carrier-free, i.e. 100% pure. Data for ^{125}I and ^{131}I are given in Table 2.

Carrier-free ^{131}I is now available commercially but, in the early days of RIA, this was not the case with the result that ^{131}I-labelled radioligands had about half the count rate of their ^{125}I-labelled counterparts[79]. This and the reasons already mentioned encouraged the adoption of ^{125}I, and nowadays ^{131}I, ^{123}I (half-life 13.3 hours) and other short-lived isotopes are mainly employed in diagnostic nuclear medicine while ^{125}I is virtually the only γ-emitting isotope used in RIA.

4.2.1 ^{125}I-Labelled Drugs

Few ^{125}I-labelled drug derivatives are available commercially so they must either be synthesised "in-house" or else RIA kits must be used. Radioiodination methods have been reviewed[80, 81].

4.2.2 Oxidative Radioiodination Methods

The chloramine T method[82, 83], the most widely used radioiodination technique, was developed for labelling proteins but has been widely applied to drugs or their derivatives containing a phenol or imidazole ring. ^{125}I as Na^{125}I (approx. 37 MBq) is added to a solution of the substrate (ng-µg quantity) followed by a solution of chloramine T (µg-mg/ml) which is the sodium salt of N-monochloro-p-toluenesulphonamide. The reaction is virtually instantaneous and is stopped by the addition of a slight excess of sodium metabisulphite. Chloramine T forms HOCl in solution which oxidises the radioiodide to give an electrophilic species[84, 85], possibly $H_2O^{125}I^+$. Electrophilic substitution of ^{125}I into the phenol or imidazole ring then occurs. The reaction is generally carried out in aqueous solution at pH7. The efficiency is reduced at higher or lower pH values and, at a pH greater than 8.5, substitution into imidazole rather than phenolic rings occurs. The reactions are shown below:—

R–C$_6$H$_4$–OH or 4-R-imidazole
1) Na^{125}I, Chloramine T
2) $Na_2S_2O_5$
→ R–C$_6$H$_3$(^{125}I)–OH or 4-R-2-^{125}I-imidazole

Substitution of more than one ^{125}I into the rings occurs to a varying extent depending on the reaction conditions. Mono[^{125}I]iodo-histamine is more stable than the di- or tri-substituted compounds[86] but the stabilities of radioiodinated phenols do not appear to have been studied.

Table 2. Characteristics of ^{125}I and ^{131}I

Isotope	Half-life (days)	Specific activity at 100% isotopic abundance (TBq/milligramatom)	Photon energy (γ, MeV)
^{131}I	8.04	559	0.080 – 0.723
^{125}I	60.0	80.7	0.035

Ring systems other than phenols and imidazoles that are subject to electrophilic attack can be radioiodinated by the chloramine T method[80, 87] but this possibility has been somewhat neglected in drug RIA.

After radioiodination, the radioligand and any unreacted drug can be separated from unreacted $^{125}I^-$ by ion exchange, solvent extraction or chromatography. In theory, $^{125}I^-$ can be precipitated virtually quantitatively as $Ag^{125}I$ but, in practice using $AgNO_3$, HNO_3 and KI (as a carrier), the efficiency was found to be lower than expected and the radioligand that was prepared ([^{125}I]iodo-*p*-hydroxybenzoylecgonine) was strongly adsorbed by the precipitated $Ag^{125}I$ and AgI[88]. A high specific activity radioligand is not always necessary in drug RIA but, if required, pure radioligand can be isolated by chromatography.

Chloramine T is not the only oxidizing agent employed in radioiodination. Chlorine, sodium hypochlorite, iodic acid, nitrous acid, nitric acid and hydrogen peroxide have also been used but not all substrates are stable under strong oxidizing conditions and so milder conditions or alternative methods are sometimes required.

Chloramine T attached to a solid phase has been marketed as "Iodobeads" (Pierce (UK) Ltd., Cambridge, U.K.). One or more beads are placed in the radioiodination mixture where they function as very dilute chloramine T. The reaction is terminated easily by removing the beads and so no bisulphite is required.

Iodogen (1,3,4,6-tetrachloro-3α,6α-diphenylglycouril)[89] is a stable, water-insoluble, mild oxidizing agent for radioiodination that can be plated on to the walls of a reaction vessel by evaporating a chloroform solution. Removal of a radioiodination mixture from the reaction vessel terminates the reaction.

Another mild, solid-phase oxidizing agent is Protag-125 which consists of glycouril bound to silica particles (J. T. Baker Chemical Co., Phillipsburgh, N.J., U.S.A.).

N-bromosuccinimide is a mild oxidizing agent that has been used to radioiodinate proteins but not, as yet, drugs[90].

Surface-catalysed oxidation of iodide to an iodinating species occurs on the silica gel of a thin-layer chromatography (TLC) plate and has been employed in the radioiodination of a variety of molecules including substituted phenols and anilines[91, 92].

Electrolysis can be used to oxidize iodide to iodine for radioiodination under mild conditions with a potential radiochemical yield of 100%. Originally applied to proteins, the method has subsequently been used to radioiodinate various small molecules[80].

Enzymatic radioiodination is another mild alternative to chemical oxidation methods[80]. Lactoperoxidase has been used with low levels of hydrogen peroxide to catalyse the oxidation of radioiodide to radioiodine. Glucose oxidase can be used to generate the hydrogen peroxide *in situ* from glucose, ensuring very mild conditions throughout the reaction. A solid-phase mixture of the enzymes is available (Bio-Rad Laboratories, Richmond, California, U.S.A.). The technique is less suitable for imidazole than phenolic rings; attempts to radioiodinate bleomycin which contains an imidazole ring resulted in preferential labelling of the enzyme[93, 94].

An early method of oxidative radioiodination of proteins was reaction with molecular radioiodine, which is now available commercially but was originally prepared by the oxidation of radioiodide[80]. Disadvantages of the method are the 50% maximum radiochemical yield (half of the radioiodine ends up as radioiodide) and the increased hazard to the operator due to the volatility of radioiodine.

The use of iodine monochloride rather than molecular iodine affords a potential radiochemical yield of 100% since iodine monochloride is polarised with a partial positive charge on the iodine[95]. Radioiodination of a phenol by molecular radioiodine or radioiodine monochloride proceeds by electrophilic attack on the phenoxide ion with the slow loss of a proton, and so the method is applicable to molecules susceptible to electrophilic attack, including electrophilic attack on a double bond[80].

4.2.3 Conjugation Labelling: Homologous and Heterologous Assays

Few drugs have structures that enable them to be radioiodinated by the chloramine T or another oxidative method. Sometimes it is feasible to synthesise a close analogue with, for instance, a phenolic hydroxyl group e.g.[96], but often a radioiodinatable moeity must first be attached to the drug, a procedure known as prosthetic group or conjugation labelling. Tyrosine methyl ester, tyramine and histamine have all been used as prosthetic groups. The prosthetic group may be radioiodinated before linking it covalently to the drug, thus avoiding exposure of the drug to oxidizing conditions. As a general rule, the prosthetic group should be linked to the drug at the position used for synthesizing the drug-protein conjugate required for raising the antiserum (Sect. 5).

The drug-prosthetic group and drug-protein linkages can be identical or non-identical and are described as "homologous" or "heterologous" respectively. A disadvantage of homologous linkages is that some antibodies may bind avidly to the bridge region of the radioligand, thus reducing the potential sensitivity of the assay. A heterologous assay is therefore preferable when high sensitivity is required, but homologous assays are often satisfactory for detecting therapeutic levels of drugs.

Heterologous assays, including the use of different stereochemical configurations in the antigen and radioligand[97], have been applied to steroids since high sensitivity is necessary to detect the low levels of these hormones that are found in blood[80, 98–104]. However, a novel, high sensitivity, homologous assay for progesterone has been described[105, 106] in which the bridge was the glucuronide moeity of 11α-progesterone-1β-glucuronide. It was postulated that the glucuronide bridge was only weakly immunogenic, thus enabling a sensitive homologous assay to be developed. In contrast, a highly specific pregnanediol-3α-glucuronide antiserum was raised using an immunogen in which the steroid was linked to protein via the steroid nucleus, leaving the glucuronide exposed[107].

High-sensitivity homologous assays have also been developed by using a short bridge that closely resembles the structure of the analyte[108] and by using monoclonal antibodies (Sect. 5.6) that do not recognise the bridge[109].

Bridge recognition in a homologous assay has been used as a means of increasing sensitivity in the RIA of 5-hydroxyindole acetic acid[110]. The analyte is converted to an amide derivative which better resembles the radioligand and immunizing conjugate, thus increasing the sensitivity of the assay 100-fold. This strategy was inspired by previous work on cyclic nucleotides[111].

Synthetic methods for attaching prosthetic groups to drugs or suitable derivatives thereof are the same as those used to prepare drug-protein immunogens (Sect. 5.1).

Radioiodinated prosthetic groups that condense spontaneously with primary amines are used widely for labelling proteins and to a limited extent for labelling drugs. The best-known example of these is the Bolton-Hunter reagent[112], *N*-succinimidyl 3-(4-hydroxy-5-[^{125}I]iodophenyl)propionate, which has been used to prepare radioligands from drugs such as clonazepam[113], desipramine[114] and bleomycin[94]. Wood's reagent[115], methyl-3,5-di[^{125}I]iodohydroxybenzimidate, is similar but preserves the charge on the amino group with which it condenses. Both reagents are commercially available. Other reagents which serve the same purpose are 4-[^{125}I]iodobenzenediazonium chloride[116], diazotized 3-[^{125}I]iodo-4-aminobenzenesulphonic acid[117, 118], *tert*-butyloxycarbonyl-L-[^{125}I]iodotyrosine-*N*-hydroxysuccinimide ester[119], [^{125}I]diiodofluorescein isothiocyanate[120] and 7-hydroxy-[^{125}I]iodocoumarin-3-acetic acid-*N*-hydroxysuccinimidyl ester[121, 122] (7-hydroxy-coumarin acetic acid is also known as 4-methylumbelliferyl-3-acetic acid). The fluorescein and coumarin derivatives can also be used as fluorescent labels and so have a dual function.

Reagents that react with carbonyl rather than amino groups include 3-[^{131}I]iodopyridinium hydrazonyl acetate[123], 3-(3-[^{125}I]iodo-4-hydroxyphenyl)propionyl carbohydrazide[124] and 3-[^{125}I]iodobenzoyl hydrazide[125].

Heterobifunctional reagents (Sect. 5.1.1) that can be radioiodinated have been described[126] but have yet to be used to prepare radioligands for RIA.

4.2.4 Other Radioiodination Methods

Oxidative or prosthetic group radioiodinations are widely employed for radioligand synthesis. A lesser used alternative is the substitution of radioiodine for stable iodine or other suitable atom or group. Various substitution methods have been devised and applied to a variety of small molecules but few drugs[80]. Compounds containing diazo, triazine or silane groups, stable iodine, bromine, boron, tin or thallium are all amenable to substitution with radioiodine. With stable iodine as the leaving group, the specific activity of the product will be low, but other leaving groups enable high specific activities to be attained. The potential scope of the method is considerable since a suitable leaving group can be incorporated into almost any molecule.

Little used radioiodination methods include heating the compound to be labelled in a melt with Na^{125}I, and recoil labelling in which highly positive radioiodine species produced by radioxenon decay react with the substrate[80]. The melt procedure is simple but its mechanism is unknown, while recoil labelling requires access to a cyclotron.

4.2.5 Stability and Storage of ^{125}I-Labelled Drugs

Compounds labelled with ^{125}I are more stable than those labelled with ^{131}I since ^{125}I does not emit β-particles when it decays. In general, ^{125}I-labelled radioligands are best stored in methanol or ethanol solution at low temperature, although some compounds are remarkably stable in aqueous buffer. After two half-lives (approximately four months) the count rate is likely to be too low for RIA and the synthesis will have to be repeated. This is only a minor disadvantage since the commonly-used radioiodination methods are simple, reliable, speedy and virtually hazard-free.

4.2.6 γ-Counting

γ-Counting is similar in principle to liquid scintillation counting (Sect. 4.2) but is simpler in practice. γ-rays emitted by the sample (liquid or solid) pass through the walls of the container and are absorbed by a solid scintillator or fluor consisting of a large sodium iodide crystal which is usually 5–8 cm in diameter. A cylindrical hole in the crystal forms the sample chamber. The sodium iodide contains about 0.1% thallium iodide which improves its fluorescent properties. One face of the crystal is connected to a photomultiplier tube while the remainder of the surface is coated with aluminium to exclude light and to protect it from atmospheric moisture. The energy of each γ-ray is absorbed by the crystal and reemitted as a pulse of photons in the visible region. The pulses are registered by the photomultiplier tube, the magnitude of the outputs being proportional to the energies of the γ-rays. As with liquid scintillation counting, one or more pulse height analysers can be adjusted or are preset for counting different isotopes. Most modern instruments have automatic sample changers and data processing facilities, and some have a number of matched counting wells so that a number of samples can be counted simultaneously, thus reducing the overall counting time.

γ-Counters are generally cheaper than liquid scintillation counters since only one photomultiplier tube per counting head is required and the sample changing mechanism does not require shutters to exclude light. In addition, γ-counting saves the cost of liquid scintillant, liquid scintillation tubes and the preparation of samples for counting, and avoids the need to dispose of large volumes of used liquid scintillant.

As with liquid scintillation counters, the purchaser is advised to compare prices and specifications in the light of his budget and requirements.

4.3 Specific Activity

With a radioligand of known specific activity, the amount used in an assay can be calculated and kept constant despite batchwise variations in quality. Manufacturers generally provide the necessary data for their products, and methods of determining specific activity have been published [127–129] but these require either that the labelled and unlabelled antigens are equally immunoreactive or that a non-radioactive (i.e. ^{127}I-labelled) analogue of the radioligand is available. ^{3}H-labelled drugs behave like their unlabelled counterparts and ^{125}I-labelled proteins are usually assumed to behave like unlabelled proteins, but ^{125}I-labelled and unlabelled drugs can differ greatly. Direct estimation of the specific activity of a radioiodinated drug by measuring its activity and mass is difficult but, if a chromatographically pure product can be isolated, its specific activity can be assumed to be that of ^{125}I, namely 80.7 TBq/mmol for a monoiodinated compound. Often, however, the specific activity of the label is not a particularly critical factor in drug RIA and it is necessary only to check that a new batch of label performs adequately in the assay.

5 Antiserum Production

A good antiserum is a prerequisite for a successful RIA. Commercial organisations are concerned with drugs of clinical rather than forensic importance and so there are few suppliers of antisera for drugs of abuse. Costs vary and it may prove less expensive to raise one's own antiserum or pay to have it raised than to purchase in bulk. Samples of all available antisera should be tested before purchase (Sect. 3). Estimation of the titres and cross-reactivity patterns will generally narrow the field, and likely candidates can be tested further to ensure that there are no endogenous interferences in the samples to be analysed and that the detection limits are satisfactory.

Emit enzyme immunoassay kits (Syva, Palo Alto, California, U.S.A.) are an inexpensive source of drug antisera for RIA[72, 73, 96, 114, 130–132]. The antiserum (approximately 5 ml) from a single Emit kit (nominally 100 enzyme immunoassay tubes) can be diluted thirty times or more (600 times in one instance[96]) and is thus sufficient for thousands of RIA tubes. A disadvantage is that successive batches of Emit antiserum may perform differently in RIA although they are suitable for their intended purpose.

Antiserum production depends on the ability of animals to produce antibodies in response to an enormously diverse range of antigens[133]. In order to raise an antiserum, a suitable immunogen must first be prepared. Particulate cellular material, proteins and high molecular weight polypeptides and polysaccharides are naturally immunogenic, that is, when introduced into an animal parenterally (i.e. not via the alimentary canal) they induce antibody formation. Substances with a molecular weight of less than about 5000 are less effective immunogens (angiotensin, gastrin and vasopressin with molecular weights of about 1000–2000 are exceptions) and small molecules such as drugs must be linked to a large molecule, usually a protein, to form an immunogenic conjugate.

Various methods have been devised for preparing drug-protein conjugates (Sect. 5.1). A standard procedure is to link the drug to the protein with a peptide bond, and so it is often necessary to synthesise first a drug derivative containing a carboxyl or primary amino group. A drug or other small molecule that is not immunogenic unless linked to a large molecule is termed a "hapten".

The pioneering studies of Landsteiner[134] established that an antiserum raised against a hapten-protein conjugate contains antibodies that recognise the part of the hapten furthest from the carrier protein. The specificity of the antiserum can therefore be controlled to some extent by varying the site of attachment of the hapten to the protein. For instance, a conjugate of morphine linked via the C3- or C6-position to protein produces an antiserum that binds both morphine and codeine[135, 136] (and various other opiates) but, if the linkage is via the C2-position of the phenolic ring[137] or the heterocyclic nitrogen atom[138] of morphine, the antiserum binds preferentially to morphine and cross-reacts poorly with codeine.

Work with steroids has shown that the functional group used to conjugate a hapten to a protein may retain some efficacy as an antigenic determinant, and it has also been demonstrated that conjugation via a non-functional position produces a highly specific antiserum[139].

As a general rule, if a "drug-specific" antiserum is required, structures that are characteristic of the drug but not its metabolites or analogues should be left exposed in the immunogen while, if a "general" or "class-specific" antiserum is required, the exposed structures should be characteristic of the class of drugs including metabolites.

Antibodies can distinguish between hapten enantiomers[140–151]. Immunisation with a d,l-hapten-protein conjugate gives an antiserum that binds both d- and l-hapten but, if a d- or l-hapten-protein conjugate is used, the resulting antiserum is likely to exhibit a high degree of stereospecificity. This must be taken into account in designing an assay for a drug that consists of a single enantiomer or for a racemic drug that undergoes stereoselective metabolism.

A drug antiserum also contains antibodies that recognise parts of the carrier protein and the hapten-protein bridge. The former are of no significance in RIA but the latter can adversely affect the sensitivity of the assay if an identical bridge links a prosthetic group to the drug in the radioligand (Sect. 4.2.3).

There is evidence that the type of bridge between the hapten and the protein affects the immunogenicity of the conjugate. A diphenylhydantoin-3-acetate-thyroglobulin immunogen produced an antiserum that bound a 3-N-acetamido derivate of diphenylhydantoin preferentially to diphenylhydantoin, suggesting that the antibodies recognised the conformation of the diphenylhydantoin-3-acetate moeity of the immunogen which, due to interaction between the diphenylhydantoin carbonyl groups and the carbonyl in the bridge, would differ somewhat from the conformation of diphenylhydantoin itself. In contrast, immunogens in which diphenylhydantoin was linked to bovine serum albumin or thyroglobulin with a valerate bridge gave better antisera, and it was postulated that the hydrophobic bridge repelled the polar diphenylhydantoin, thus isolating it from the protein in the immunogen[152].

Within limits, the length of the hapten-protein bridge in an immunogen does not appear to be critical, though a 2–4 carbon atom bridge probably "exposes" the hapten and enhances its recognition by the immune system of an animal. An excessive bridge length is best avoided; for instance, the inclusion of hexanoic acid in the bridge between cholic acid and bovine serum albumin increased the bridge length from three to nine carbon atoms but lowered the titre of the resulting antiserum[153].

The number of hapten molecules attached to a carrier protein is not well correlated with the properties of the antiserum. Useful antisera can be produced with hapten: protein molar ratios ranging from about 2 to 20–30 but, with a higher degree of substitution (>40), there may be a poor response on immunisation[152, 154].

Immunising an animal with a steroid hormone-protein conjugate can cause severe physiological effects due to circulating antibodies binding to endogenous hormone and effectively rendering it non-active. This problem is unlikely to arise with drug-protein conjugates but metabolism of the haptenic portion of the conjugate can lead to an unexpected response. For instance, immunisation of rabbits with a hydroxyprazepam-succinoyl-bovine serum albumin conjugate produced an antiserum that cross-reacted with oxazepam, diazepam and desmethyldiazepam in preference to prazepam and 3-hydroxyprazepam, an indication that N-dealkylation of the hydroxyprazepam moeity of the immunogen had occured *in vivo* resulting in an oxazepam-succinoyl-bovine serum albumin conjugate[155].

An immunogen need not be soluble in order to elicit a response but one that becomes insoluble during its preparation cannot be purified thoroughly and may produce an antiserum with undesirable cross-reactions.

5.1 Immunogen Preparation

Bovine serum albumin is the most widely used carrier protein for immunogen preparation since it is inexpensive and readily available. Thyroglobulin, keyhole limpet haemocyanin and various other proteins have been used, as have synthetic polypeptides and polymers. There are no definitive rules governing the choice of a carrier though synthetic polypeptides have, on occasions, proved less effective than proteins[154].

The major functional groups of bovine serum albumin to which haptens can be linked are the ε-amino groups of lysine (59 residues), the carboxyl groups of aspartic and glutamic acids (133 residues), the phenolic hydroxyl groups of tyrosine (19 residues) and the (less reactive) imidazole groups of histidine (17 residues). Other functional groups of proteins include the sulphydryl groups of cysteine, terminal amino and carboxyl groups, the hydroxyl group of serine, and the rings of histidine, tryptophan and tyrosine residues which can be linked to haptens via diazonium salts.

Various methods have been used to attach haptens to proteins depending on the available hapten functional groups. For haptens with no suitable functional groups, an analogue with a convenient group must be synthesised or possibly a metabolite can be used. The various conjugation methods applicable to common functional groups are outlined below and the reader is referred to Erlanger's reviews[154, 156].

5.1.1 Haptens with Primary Aliphatic Amino Groups

A water-soluble carbodiimide such as 1-ethyl-3-(3-dimethylaminopropyl)-carbodiimide (EDC) can be used to form peptide bonds between the primary aliphatic amino groups of a hapten and the carboxyl groups of a carrier protein. Dicyclohexylcarbodiimide (DCC) can be used in organic solvents to activate a water-insoluble hapten but the organic solvent must be water-miscible to permit reaction with the protein. The method is broadly applicable and is best carried out around pH6 and at 0 °C. Reaction times varying from minutes to days have been employed. The mechanism is complex and not fully understood. Cross-linking of the protein occurs but is usually no problem provided the desired reaction with the hapten occurs. Dicyclohexylurea, formed from DCC, is very insoluble and may be removed by centrifugation or filtration. The urea formed from EDC is water-soluble and is separated from the immunogen by gel filtration or dialysis.

$$\underset{\text{Carbodiimide}}{R{-}N{=}C{=}N{-}R} + R'{-}NH_2 + R''{-}COOH \xrightarrow[\text{intermediates}]{\text{via}} R'{-}NHCO{-}R'' + \underset{\text{Substituted urea}}{R{-}NHCONH{-}R}$$

Carbodiimides can also react with alcohols, thiols and phenols[157].

N,N'-carbonyldiimidazole is another reagent that forms peptide bonds.

$$\underset{\textit{N,N'}\text{-Carbonyldiimidazole}}{\text{Im}{-}CO{-}\text{Im}} \xrightarrow{RCOOH} RCO{-}\text{Im} \xrightarrow{R'NH_2} RCONHR'$$

p-Nitrobenzoyl chloride reacts with primary aliphatic amines to give a *p*-nitrobenzoylamide which can be reduced to the corresponding p-aminobenzoyl compound. This can be diazotized and coupled to a protein.

$$R{-}NH_2 \xrightarrow[p\text{-Nitrobenzoyl chloride}]{p\text{-}NO_2C_6H_4COCl} p\text{-}NO_2C_6H_4CONHR \xrightarrow{\text{Reduction}} p\text{-}NH_2C_6H_4CONHR \xrightarrow{\text{Diazotization}} p\text{-}N_2^+Cl^-C_6H_4CONHR \xrightarrow[\text{(Tyrosyl resiue)}]{\text{Protein}{-}C_6H_4{-}OH}$$

$$RHNOC{-}C_6H_4{-}N{=}N{-}C_6H_3(OH){-}\text{Protein}$$

Also reacts with histidyl and tryptophyl residues of protein

Reaction with succinic anhydride[158] or with terephthalaldehydic acid followed by sodium cyanoborohydride[159] converts a primary aliphatic amine to a carboxyl group which can be linked to a protein by various methods (Sect. 5.1.5).

$$R{-}NH_2 \xrightarrow[\text{Succinic anhydride}]{} R{-}NH{-}CO{-}CH_2{-}CH_2{-}COOH$$

$$R{-}NH_2 \xrightarrow[2)\ NaBH_3CN]{1)\ OHC{-}C_6H_4{-}COOH\ \text{Terephthalaldehydic acid}} R{-}NH{-}CH_2{-}C_6H_4{-}COOH$$

More than forty bifunctional reagents (compounds with two reactive groups) are available (Pierce Chemical Company, Rockford, Illinois, U.S.A.). They are used to cross-link suitable compounds and the majority are designed to react with primary amino groups. The reactive groups of bifunctional reagents need not be identical. In recent years, bifunctional reagents have been widely used in molecular biology and to prepare enzyme-labelled compounds for immunoassays[26, 157, 160–165], but they have yet to be used to any extent for the preparation of drug-protein immunogens. The reactions of the common functional groups of bifunctional reagents are outlined in Table 3.

5.1.2 Haptens with Secondary Aliphatic Amino Groups

Secondary as well as primary aliphatic amino groups are amenable to carbodiimide condensation[171] (Sec. 5.1.1). Alternatively they may be derivatised to give other functional groups that can be linked to carrier proteins by standard methods.

Table 3.

$$R{-}N{=}C{=}O \xrightarrow{R'{-}NH_2} R{-}NH{-}\overset{\displaystyle O}{\overset{\|}{C}}{-}NH{-}R'$$

Isocyanate (also reacts with hydroxyl and thiol groups[157])

$$R{-}N{=}C{=}S \xrightarrow{R'{-}NH_2} R{-}NH{-}\overset{\displaystyle S}{\overset{\|}{C}}{-}NH{-}R'$$

Isothiocyanate (less reactive than isocyanate. Also reacts with secondary amines[166])

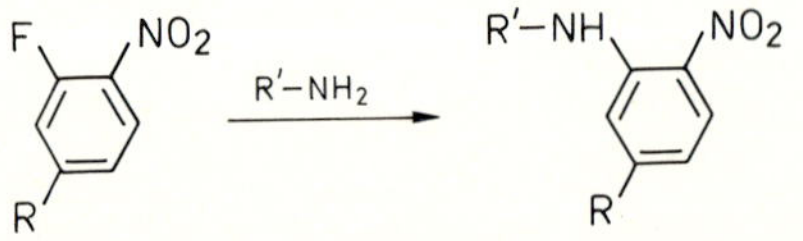

$$\text{F–}C_6H_3(NO_2)\text{–R} \xrightarrow{R'{-}NH_2} R'{-}NH{-}C_6H_3(NO_2){-}R$$

Aryl halide (also reacts with phenolic hydroxyl, thiol and imidazole groups)

$$R{-}\overset{\displaystyle O}{\overset{\|}{C}}{-}CH_2Br \xrightarrow{R'{-}NH_2} R{-}\overset{\displaystyle O}{\overset{\|}{C}}{-}CH_2{-}NH{-}R'$$

Active halogen (also reacts with thiol, sulphide and imidazole groups)

$$R{-}\overset{\displaystyle NH}{\overset{\|}{C}}{-}OCH_3 \xrightarrow{R'{-}NH_2} R{-}\overset{\displaystyle NH}{\overset{\|}{C}}{-}NH{-}R'$$

Imidoester

$$R{-}\overset{\displaystyle O}{\overset{\|}{C}}{-}O{-}N(\text{succinimide}) \xrightarrow{R'{-}NH_2} R{-}\overset{\displaystyle O}{\overset{\|}{C}}{-}NH{-}R'$$

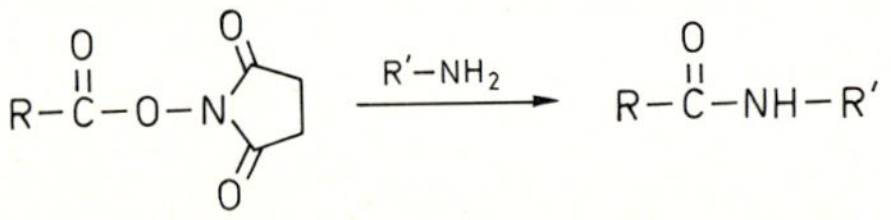

N-Hydroxysuccinimide ester

$$R{-}\overset{\displaystyle O}{\overset{\|}{C}}{-}Cl \xrightarrow{R'{-}NH_2} R{-}\overset{\displaystyle O}{\overset{\|}{C}}{-}NH{-}R'$$

Acid chloride (also reacts with hydroxyl and secondary but not tertiary amine groups)

$$R{-}CHO \xrightarrow[\text{Reduction}]{R'{-}NH_2 \atop \text{Schiff base formation}} R{-}CH_2{-}NH{-}R'$$

Aldehyde (with glutaraldehyde, the reaction is more complicated than that depicted[157]
Thiol and, to some extent, phenolic hydroxyl and histidyl groups also react)

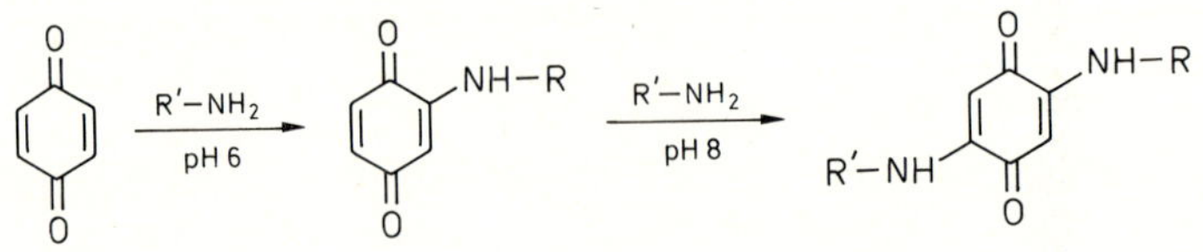

$$p\text{-benzoquinone} \xrightarrow[\text{pH 6}]{R'{-}NH_2} \text{2-(NH–R)-benzoquinone} \xrightarrow[\text{pH 8}]{R'{-}NH_2} \text{2-(NH–R)-5-(R'–NH)-benzoquinone}$$

p-Benzoquinone[167] (also reacts with thiol groups and carbohydrates)

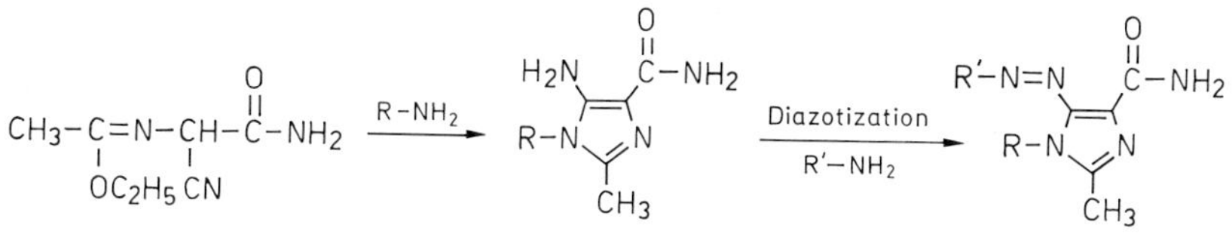

$CH_3-C(OC_2H_5)=N-CH(CN)-C(=O)-NH_2$ —R-NH$_2$→ —Diazotization, R′-NH$_2$→

Ethyl *N*-(carbamoylcyanomethyl) acetimidate[168]

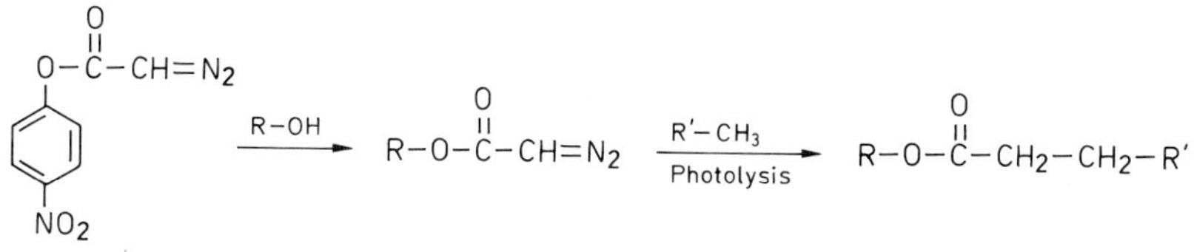

—R-OH→ $R-O-C(=O)-CH=N_2$ —R′-CH$_3$, Photolysis→ $R-O-C(=O)-CH_2-CH_2-R'$

p-Nitrophenyldiazoacetate[169]

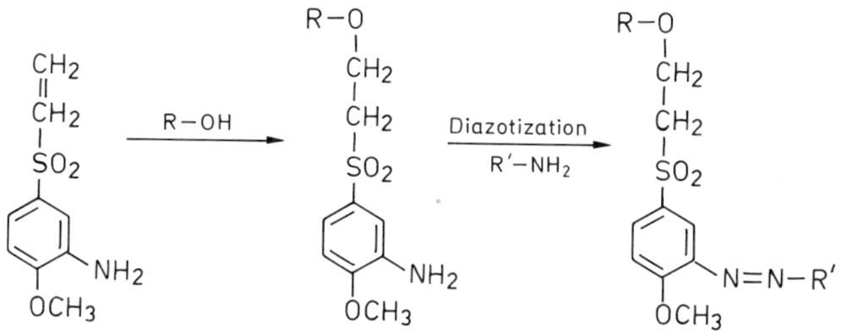

—R-OH→ —Diazotization, R′-NH$_2$→

3-Amino-4-methoxyphenyl vinyl sulphone[170]

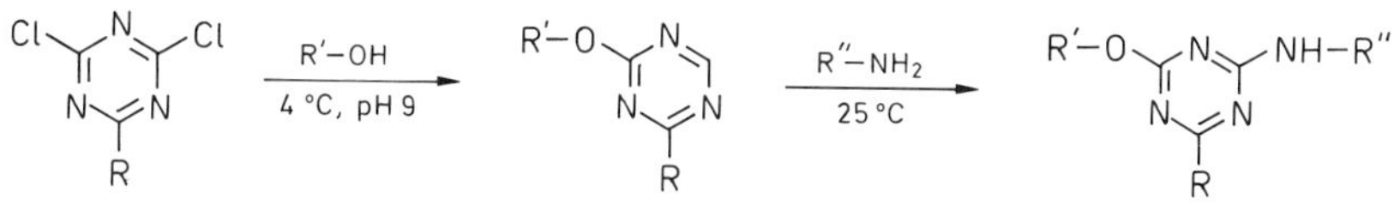

—R′-OH, 4 °C, pH 9→ —R″-NH$_2$, 25 °C→

Triazine derivative

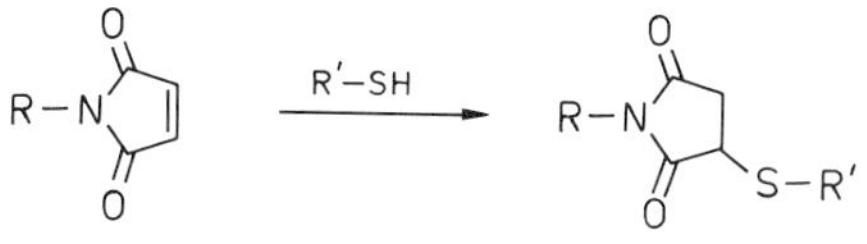

—R′-SH→

Maleimide

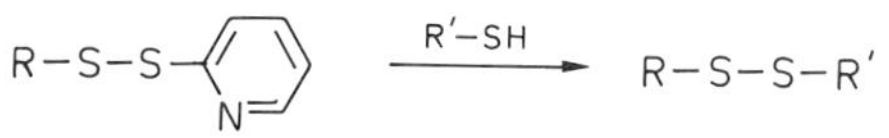

—R′-SH→ R-S-S-R′

2-Dithiopyridyl

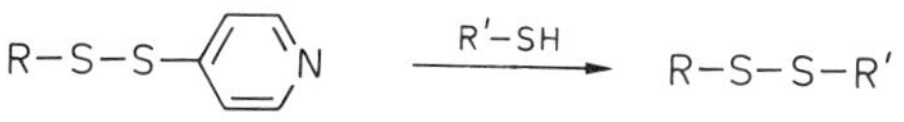

—R′-SH→ R-S-S-R′

4-Dithiopyridyl

—Photolysis, Protein→

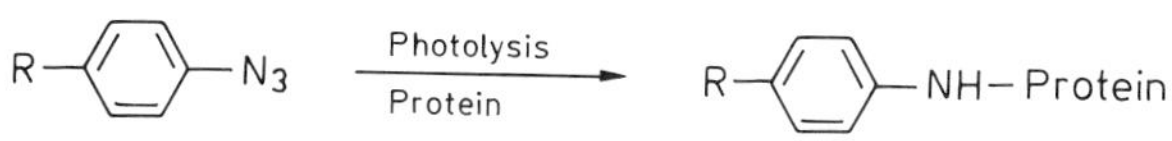

Azidophenyl (non-selective reaction)

Reaction with *N*-(4-bromobutyl)phthalimide followed by hydrolysis adds a 4-carbon chain terminating in a primary amino group[172].

$$\mathrm{RR'NH} \xrightarrow[\textit{N}\text{-(4-Bromobutyl) phthalimide}]{\mathrm{Br{-}CH_2{-}CH_2{-}CH_2{-}CH_2{-}N(phthalimide)}} \mathrm{RR'N{-}CH_2{-}CH_2{-}CH_2{-}CH_2{-}N(phthalimide)}$$

$$\xrightarrow{\text{Hydrolysis}} \mathrm{RR'N{-}CH_2{-}CH_2{-}CH_2{-}CH_2{-}NH_2}$$

Reaction with sodium chloroacetate under basic conditions gives a carboxyl group[173].

$$\mathrm{RR'NH} \xrightarrow[\text{Sodium chloroacetate}]{\mathrm{ClCH_2{-}COONa}} \mathrm{RR'N{-}CH_2{-}COOH}$$

Reaction with ethylchloroacetate[152] or ethylbromoacetate[174] followed by hydrolysis gives a carboxyl group. Ethyl-3-bromopropionate[175] and methyl-5-bromovalerate[176] react similarly and yield longer side-chains.

$$\mathrm{RR'NH} \xrightarrow[\text{Ethyl chloroacetate}]{\mathrm{ClCH_2{-}COOC_2H_5}} \mathrm{RR'NH{-}CH_2{-}COOH}$$

Reaction with succinic anhydride[177] gives a carboxyl group.

$$\mathrm{RR'NH} \xrightarrow[\text{Succinic anhydride}]{} \mathrm{RR'N{-}\overset{O}{\overset{\|}{C}}{-}CH_2{-}CH_2{-}COOH}$$

Michael addition with methyl acrylate[177] gives a carboxyl group.

$$\mathrm{RR'NH} \xrightarrow[\text{2) Hydrolysis}]{\text{1) } \mathrm{CH_2{=}CH{-}COOCH_3}\text{ Methyl acrylate}} \mathrm{RR'N{-}CH_2{-}CH_2{-}COOH}$$

Carboxyl groups can be linked to proteins by methods given in Sect. 5.1.5.

5.1.3 Haptens with Aromatic Amino Groups

Reaction with nitrous acid gives a diazonium salt which reacts with tyrosyl, tryptophyl and histidyl residues of proteins.

$$\mathrm{R{-}C_6H_4{-}NH_2} \xrightarrow[\text{acid}]{\text{Nitrous}} \mathrm{R{-}C_6H_4{-}N{=}N^{\oplus}} \xrightarrow[\text{(Tyrosyl residue)}]{\text{Protein}{-}\mathrm{C_6H_4{-}OH}} \mathrm{R{-}C_6H_4{-}N{=}N{-}C_6H_3(OH){-}Protein}$$

Reaction of the diazotised hapten with methyl-*p*-hydroxybenzimidate or methyl-3,5-dihydroxybenzimidate gives a product that condenses spontaneously with protein amino groups[164].

R — $N{=}N^{\oplus}$ —(HO–C₆H₄–C(=NH)–O–CH₃, Methyl-*p*-hydroxybenzimidate)→ R — N=N — HO–C₆H₃–C(=NH)–O–CH₃ —(Protein-NH_2)→ R — N=N — HO–C₆H₃–C(=NH)–NH–Protein

Reaction with phosgene converts the amino group to an isocyanate which condenses spontaneously with protein amino groups[178, 179].

R — NH_2 —($COCl_2$, Phosgene)→ R — N=C=O —(Protein-NH_2)→ R — NH—C(=O)—NH—Protein

Reaction with succinic[180] or glutaric [181] anhydride gives a carboxyl group that can be linked to proteins by methods given in Sect. 5.1.5.

R — NH_2 —(Succinic anhydride)→ R — NH–C(=O)–CH_2–CH_2–COOH

5.1.4 Haptens with Nitro Groups

Haptens containing nitro groups can be reduced to the corresponding amino compounds and linked to proteins by the methods described in Sect. 5.1.1 and 5.1.3.

5.1.5 Haptens with Carboxyl Groups

The mixed anhydride procedure, which was originally developed for peptide synthesis, is widely used to couple carboxyl-containing haptens to carrier proteins[154]. Reaction of a carboxyl group with an alkylchlorocarbonate such as isobutylchloroformate under basic conditions gives a mixed anhydride. Subsequent reaction with a protein results in acylation of the protein amino groups with the formation of peptide bonds.

R–COOH —($(CH_3)_2$CH–CH_2–O–C(=O)–Cl, Isobutylchloroformate)→ R–C(=O)–O–C(=O)–CH_2–CH(CH_3)$_2$ —(Protein-NH_2)→ R–C(=O)–NH–Protein

Carbodiimides, *N,N'*-carbonyldiimidazole and numerous bifunctional reagents can be used to link carboxyl and amino groups (Sect. 5.1.1 and Table 3). 1-Ethoxycarbonyl-2-ethoxy-1,2-dihydroquinoline has been used similarly[182].

N-hydroxysuccinimide esters, which condense spontaneously with amino groups, can be prepared from carboxyl-containing haptens and *N*-hydroxysuccinimide using a carbodiimide. The procedure was introduced when direct carbodiimide condensation of insect juvenile hormone carboxyl groups and protein amino groups failed[154].

$$\text{R-COOH} + \text{HO-N(succinimide)} \xrightarrow{\text{Carbodiimide}} \text{R-}\overset{\text{O}}{\overset{\|}{\text{C}}}\text{-O-N(succinimide)} \xrightarrow{\text{Protein-NH}_2} \text{R-}\overset{\text{O}}{\overset{\|}{\text{C}}}\text{-NH-Protein}$$

When it was found that meperidinic acid could not be coupled directly to protein, a side chain was added by preparing the acid chloride, reacting this with ethyl glycollate and hydrolysing the product to give a terminal carboxyl group that was amenable to the mixed anhydride procedure[183].

$$\text{R-COOH} \xrightarrow[\text{Thionyl chloride}]{\text{SOCl}_2} \text{R-COCl} \xrightarrow[\text{Ethyl glycollate}]{\text{HO-CH}_2\text{-}\overset{\text{O}}{\overset{\|}{\text{C}}}\text{-O-CH}_2\text{-CH}_3} \text{R-}\overset{\text{O}}{\overset{\|}{\text{C}}}\text{-O-CH}_2\text{-}\overset{\text{O}}{\overset{\|}{\text{C}}}\text{-O-CH}_2\text{-CH}_3$$

$$\xrightarrow{\text{Hydrolysis}} \text{R-}\overset{\text{O}}{\overset{\|}{\text{C}}}\text{-O-CH}_2\text{-COOH}$$

Aspirin-protein conjugates have been prepared by converting aspirin to acid chloride[184] or azide[185] derivatives that react directly with proteins.

$$\text{R-COOH} \longrightarrow \text{R-COCl} \xrightarrow{\text{Protein-NH}_2} \text{R-}\overset{\text{O}}{\overset{\|}{\text{C}}}\text{-NH-Protein}$$

$$\text{R-COCl} \downarrow \text{R-CON}_3 \xrightarrow{\text{Protein-NH}_2} \text{R-}\overset{\text{O}}{\overset{\|}{\text{C}}}\text{-NH-Protein}$$

5.1.6 Haptens with Aliphatic Hydroxyl Groups

Immunogens are usually prepared from hydroxyl-containing haptens by the introduction of a carboxyl group and conjugation to a protein by the methods described in Sect. 5.1.5.

Reaction with a dicarboxylic acid anhydride to give a half ester is widely employed[154].

$$\text{R-OH} \xrightarrow[\text{Succinic anhydride}]{} \underset{\text{Hemisuccinate}}{\text{R-O-}\overset{\text{O}}{\overset{\|}{\text{C}}}\text{-CH}_2\text{-CH}_2\text{-COOH}}$$

Reaction with a dicarboxylic acid chloride followed by hydrolysis also gives a half ester[186].

$$\text{R-OH} \xrightarrow[\text{Hydrolysis}]{\text{Cl-}\overset{\text{O}}{\overset{\|}{\text{C}}}\text{-CH}_2\text{-CH}_2\text{-}\overset{\text{O}}{\overset{\|}{\text{C}}}\text{-Cl}\ \text{Succinoyl dichloride}} \underset{\text{Hemisuccinate}}{\text{R-O-}\overset{\text{O}}{\overset{\|}{\text{C}}}\text{-CH}_2\text{-CH}_2\text{-COOH}}$$

Oxidation of hydroxyl to carboxyl has been applied to nucleosides but not, apparently, to drugs[154].

Reaction with an equimolar amount of phosgene yields a chlorocarbonate which reacts directly with protein amino groups (Schotten-Baumann reaction).

$$R{-}OH \xrightarrow[\text{Phosgene}]{COCl_2} R{-}O{-}\overset{\overset{O}{\|}}{C}{-}Cl \xrightarrow{\text{Protein-}NH_2} R{-}O{-}\overset{\overset{O}{\|}}{C}{-}NH{-}\text{Protein}$$

N,N′-carbonyldiimidazole can be used to link hydroxyl groups to protein amino groups[187].

$$R{-}OH \xrightarrow[\text{N,N'-Carbonyl-diimidazole}]{} R{-}O{-}\overset{\overset{O}{\|}}{C}{-}N\langle\text{imidazole}\rangle \xrightarrow{\text{Protein-}NH_2} R{-}O{-}\overset{\overset{O}{\|}}{C}{-}NH{-}\text{Protein}$$

Imidazoyl carbamate

Sebacoyl dichloride, a bifunctional reagent (Table 3), can be used to link hydroxyl groups to proteins[154].

Methods for linking carbohydrates to proteins may find limited use in preparing drug immunogens. Such methods[154] include:

(i) Periodate oxidation of vicinal hydroxyl groups to a dialdehyde. This reacts with a protein amino group to form an aldimine which is stabilised by reduction with borohydride or cyanoborohydride.

(ii) Conversion to a *p*-aminophenylglycoside and attachment to a protein via diazotisation or the formation of an isothiocyanate derivative.

(iii) Condensation of vicinal hydroxyl groups with laevulinic acid ethyl ester followed by saponification to give a terminal carboxyl group[188] which can be linked to a protein by the methods described in Sect. 5.1.5.

(iv) Reductive amination with cyanoborohydride to link the aldehyde form of the carbohydrate directly to protein amino groups[189] (see also Sect. 5.1.8).

(v) Reaction with β-(*p*-aminophenyl)ethylamine to form an *N*-alkylglycoside which is reduced with borohydride to a stable secondary amine[190]. The resulting carbohydrate-phenethylamine derivative is then linked to a protein via diazotisation[191] or formation of an isothiocyanate[192].

5.1.7 Haptens with Phenolic Hydroxyl Groups

An *O*-carboxymethyl group can be introduced by reaction with sodium β-chloroacetate[193]. Bromoacetate and bromopropionate have been used similarly[175]. The resulting carboxyl group can be linked to a protein by the methods desccribed in Sect. 5.1.5.

$$R{-}C_6H_4{-}OH \xrightarrow[\text{Sodium } \beta\text{-chloroacetate}]{CH_2Cl{-}COONa} R{-}C_6H_4{-}O{-}CH_2{-}COOH$$

Reaction with nitric acid introduces a nitro group which can be reduced and linked to a protein (Sect. 5.1.4).

5.1.8 Haptens with Carbonyl Groups

Aldehydes and ketones react with *O*-(carboxymethyl)hydroxylamine to give *O*-(carboxymethyl)oxime derivatives whose carboxyl groups can be linked to a protein as described in Sect. 5.1.5[154].

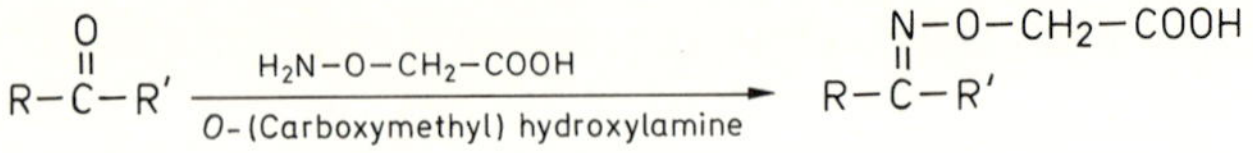

$$R-\overset{O}{\overset{\|}{C}}-R' \xrightarrow[\text{O-(Carboxymethyl) hydroxylamine}]{H_2N-O-CH_2-COOH} R-\overset{N-O-CH_2-COOH}{\overset{\|}{C}}-R'$$

Reaction with a phenylhydrazine derivative, *p*-hydrazinobenzoic acid, serves the same purpose[194].

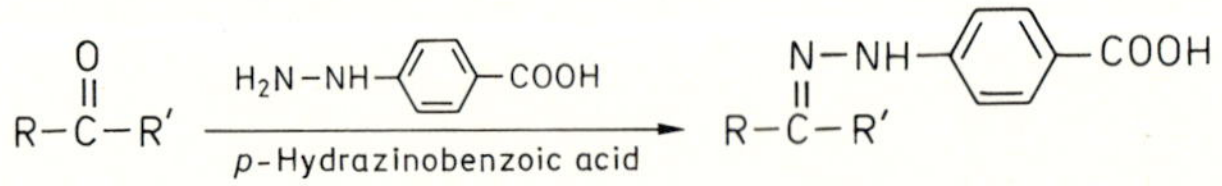

$$R-\overset{O}{\overset{\|}{C}}-R' \xrightarrow[\text{p-Hydrazinobenzoic acid}]{H_2N-NH-C_6H_4-COOH} R-\overset{N-NH-C_6H_4-COOH}{\overset{\|}{C}}-R'$$

An interesting variation is the reaction of colchicine with ethylenediamine[195]. The resulting primary amino group was linked to a carrier protein with glutaraldehyde.

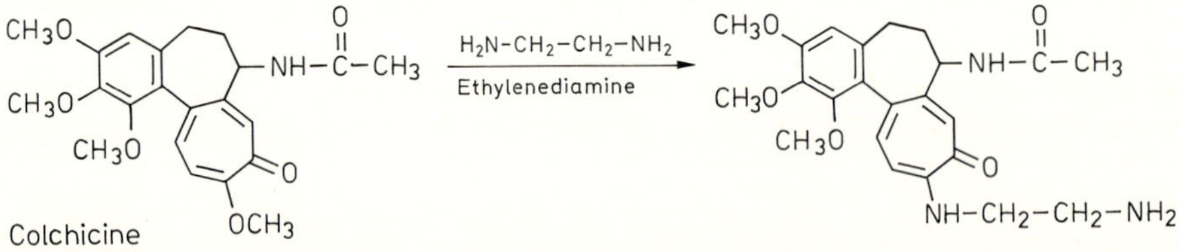

Colchicine $\xrightarrow[\text{Ethylenediamine}]{H_2N-CH_2-CH_2-NH_2}$ (OCH₃ of the tropolone ring replaced by $NH-CH_2-CH_2-NH_2$)

Aldehydes can be linked directly to protein amino groups by Schiff base formation followed by borohydride reduction to stabilize the bond[196, 197].

$$R-CHO \xrightarrow{\text{Protein-}NH_2} R-CH{=}N-\text{Protein} \xrightarrow{BH_4^{\ominus}} R-CH_2-NH-\text{Protein}$$

Aldehydes and ketones also participate in the Mannich reaction (Sect. 5.1.9).

5.1.9 Haptens with Active Hydrogen

Haptens with an active hydrogen atom can be linked to protein amino groups by the Mannich reaction which is the condensation of an active methylene compound with formaldehyde and an amine (ammonia and primary and secondary amines react)[154].

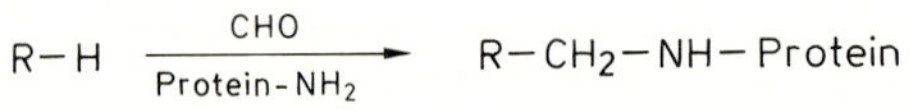

$$R-H \xrightarrow[\text{Protein-}NH_2]{CHO} R-CH_2-NH-\text{Protein}$$

An example is the linkage of tiamenidine to a carrier protein[198].

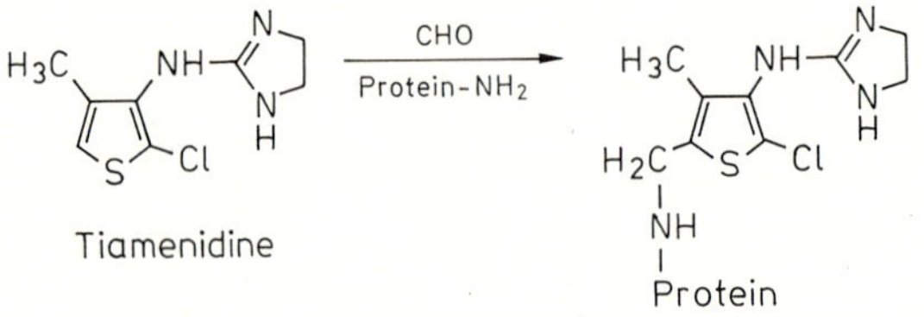

Tiamenidine $\xrightarrow[\text{Protein-}NH_2]{CHO}$ (thiophene ring substituted with $H_2C-NH-\text{Protein}$)

A carboxyl group can be attached to an active site of an aromatic ring by reaction with diazotised *p*-aminobenzoic acid. Substituent groups can influence the position at which diazo coupling occurs. The carboxyl group can then be linked to protein as described in Sect. 5.1.5.

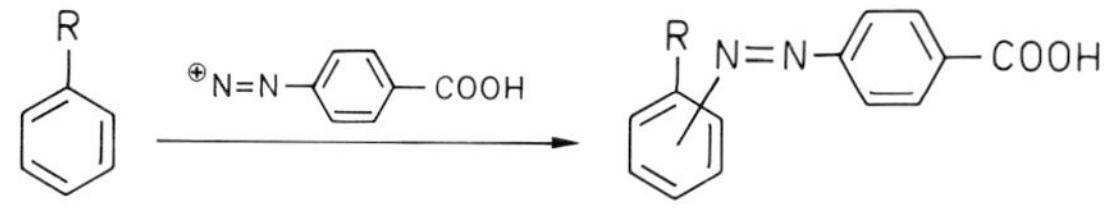

R ⊕N=N–C6H4–COOH → R N=N–C6H4–COOH

Friedel-Crafts acylation serves the same purpose and has been used to attach a carboxyl group to chlorpromazine[177].

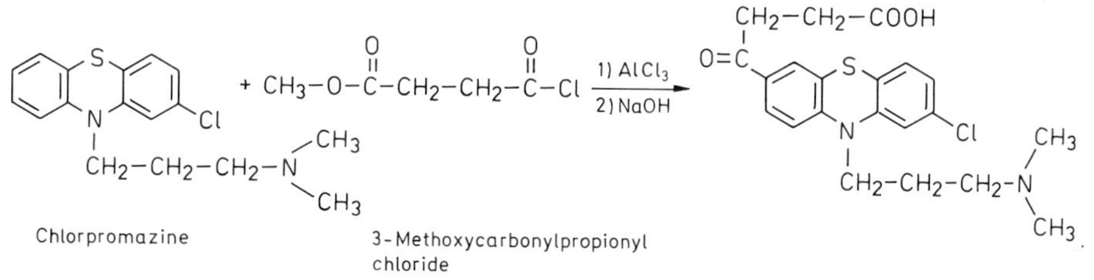

Chlorpromazine + 3-Methoxycarbonylpropionyl chloride ($CH_3-O-CO-CH_2-CH_2-CO-Cl$) → 1) $AlCl_3$ 2) NaOH → CH_2-CH_2-COOH, O=C, S, N, Cl, $CH_2-CH_2-CH_2-N(CH_3)_2$

5.1.10 Haptens with Ester Linkages

Hydrolysis of an ester is an obvious way of obtaining a useful functional group.

An alternative is exemplified by the reaction of meperidine with hydrazine to give an acid hydrazide that was diazotised to an azide and linked directly to a carrier protein[199].

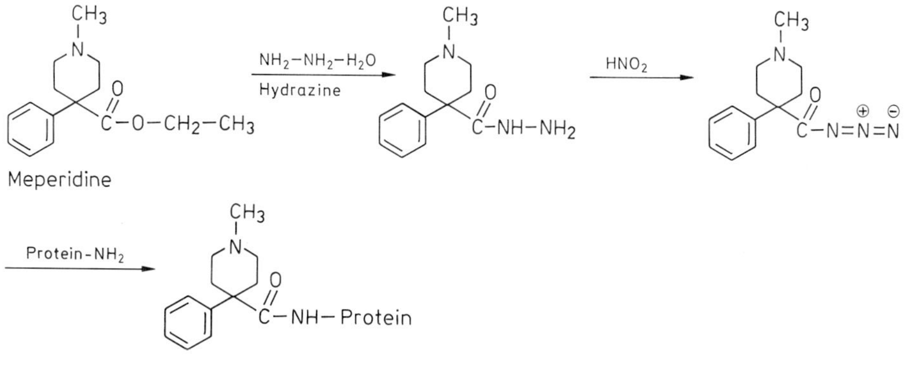

Meperidine ($C-O-CH_2-CH_3$) → $NH_2-NH_2-H_2O$ Hydrazine → $C-NH-NH_2$ → HNO_2 → $C-N=\overset{\oplus}{N}=\overset{\ominus}{N}$

→ Protein-NH_2 → C–NH–Protein

5.1.11 Haptens with Lactone Rings

Digoxin has been linked to methylated bovine serum albumin by ozonolysis of its lactone ring followed by reduction with dimethyl sulphide to give an aldehyde thought to have the structure shown below[200]. The reaction mixture after reductive ozonolysis was added to the methylated protein and the resulting linkages (presumably arising from Schiff base formation) were stabilised by reduction with cyanoborohydride (see Sect. 5.1.8).

5.1.12 Haptens with Thiol Groups

Penicillenic acid is an example of such a hapten. It has been linked to thiol-enriched protein by disulphide bond formation at pH9 in the presence of urea and hydrogen peroxide[201].

Appropriate bifunctional reagents (Table 3) offer alternative procedures for linkage via a thiol group.

5.1.13 Miscellaneous Haptens

Immunogens prepared from hydrophobic haptens may be poorly water soluble and ineffective when injected into an animal. In the case of daunamycin[165], this difficulty was avoided by coupling daunamycin to mercaptosuccinylated bovine serum albumin (which is highly water soluble) with a hydrophilic heterobifunctional cross-linking reagent, *N*-(γ-maleimidobutyryloxy)succinimide (GMBS).

Daunamycin-NH_2 Protein-NH_2

$N-(CH_2)_3-CO-O-N$ (GMBS)   1) $S-CO-CH_3$ 2) NH_2OH

Daunamycin$-NH-CO-(CH_2)_3-N$ + Protein$-NH-CO-CH_2-CH(SH)-COOH$

Daunamycin$-NH-CO-(CH_2)_3-N$ / S / $CH-CH_2-CO-NH-$Protein / COOH

Some of the methods used to prepare steroid-protein immunogens are described in the following paragraphs. Such methods may have application in the wider context of drug immunoassay. In each instance, the carboxyl derivative was linked to the carrier protein by a carbodiimide (Sect. 5.1.5).

Δ^4-3-Oxosteroids can be conjugated at the 7-position by nucleophilic addition to the 4,6-dienes[202]. Mercaptoacetic and β-mercaptopropionic acids were used since the mercapto group is an efficient nucleophile and the resulting carboxyl group is easily attached to a protein.

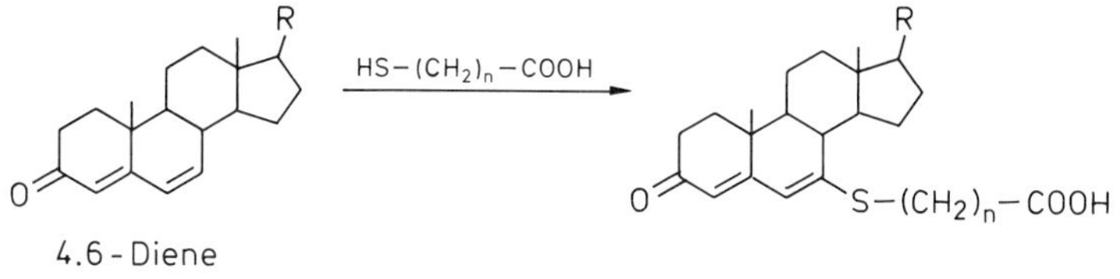

R
HS–(CH2)n–COOH
R
O
O
S–(CH2)n–COOH
4.6 - Diene

The benzylic nature of the 6-position in the oestrogens and its allylic nature in the progestins and androgens have been exploited for immunogen preparation[139].

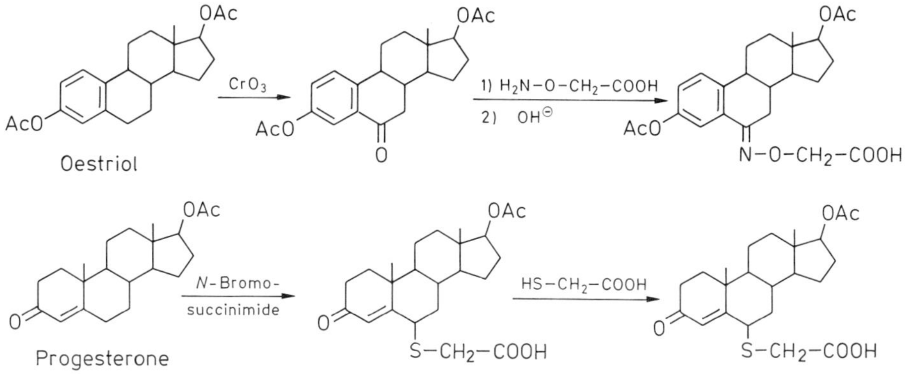

OAc
CrO3
AcO
Oestriol
OAc
1) H2N–O–CH2–COOH
2) OH⊖
AcO
O
OAc
AcO
N–O–CH2–COOH
OAc
N-Bromo-
succinimide
O
Progesterone
OAc
HS–CH2–COOH
O
S–CH2–COOH
OAc
O
S–CH2–COOH

5.2 Purification of Immunogens and Estimation of the Degree of Conjugation

It is customary to use a vast molar excess of hapten in the preparation of an immunogen. Unbound hapten is then removed, usually by dialysis[203] although gel permeation chromatography[204] and solvent extraction[203] have also been employed.

The degree of conjugation, i.e. the molar hapten: protein ratio may be estimated in various ways. The error involved may be quite high, but this matters little as there is no clear-cut relationship between the degree of conjugation and the immunogenicity of the conjugate.

The most widely applicable and arguably the most convenient way of estimating the degree of conjugation is to use a proportion of ^{3}H- or ^{14}C-labelled hapten in the preparation of the immunogen[205]. The degree of conjugation can then be calculated from the activity of the purified immunogen.

Ultraviolet spectrophotometry can be used provided the absorbance spectrum of the hapten differs from that of the protein. The weight of hapten in the immunogen may be ignored[203] or included[206] in the calculation since it has little effect on the result.

The degree of conjugation can also be estimated by measuring the unsubstituted amino groups in the immunogen (providing the hapten is conjugated via the protein amino groups) either by dinitrophenylation or by the trinitrobenzene sulphonic acid procedure[154].

With steroid sulphate-protein immunogens, the degree of conjugation can be estimated by ion chromatography of the sulphate anions liberated by hydrolysis with hydrochloric acid[207].

Hapten-protein immunogens and unsubstituted proteins have different electrophoretic mobilities. However, attempts to calculate the degree of conjugation from electrophoretic mobilities gave apparently low results[203].

Infra-red spectrophotometry provides a qualitative means of characterising hapten-protein immunogens[203] but has not been used to estimate the degree of conjugation.

A factor that appears to have been ignored in the literature is the possible presence of non-specifically bound hapten in a hapten-protein immunogen. For instance[88], in the preparation of morphine-bovine serum albumin immunogens, two experiments were carried out in which aqueous solutions of [^{3}H]morphine sulphate (instead of 3-*O*-carboxymethylmorphine[193]) and bovine serum albumin were treated twice with 1-ethyl-3-(3-dimethylaminopropyl)carbodiimide hydrochloride (EDC) over a 48 hour interval. The apparent degrees of conjugation were 0.4 and 0.7 after three days dialysis against running tap water, and 0.25 and 0.45 after a further two days dialysis. A control in which the EDC treatment was omitted had apparent degrees of conjugation of 0.15 and 0.05 after three and five days dialysis respectively. Long counting times and background subtraction minimised the errors. An immunogen that was prepared similarly from 3-*O*-carboxymethylmorphine (trace labelled with ^{3}H) and bovine serum albumin had a degree of conjugation of 7.4. The conclusion was that up to 10% of the hapten in the conjugate was not necessarily linked via its 3-position to the protein. Possibly it was linked via its 6-hydroxyl group, but this was not demonstrated. Thus a proportion of the hapten in the immunogen could give rise to unexpected antiserum cross-reactions, a useful feature in a general assay but a disadvantage if a specific assay is intended.

Such an effect could account for the observation[139] (Sect. 5) that the functional group used to attach a steroid hemisuccinate to bovine serum albumin using EDC retains some of its function as an antigenic determinant.

Emit enzyme immunoassay kits (Syva), which were mentioned at the beginning of this section as being a useful source of drug antisera, are also a source of immunogens. As well as antisera, they contain drug-enzyme conjugates that can be purified by dialysis and innoculated into animals to produce antisera. Phenytoin[204], theophylline[208] and benzodiazepine[88] antisera have been raised successfully by this method while attempts to raise barbiturate and methadone antisera[88] failed since the titres were too low to be of practical use.

5.3 Storage of Purified Immunogens

Purified immunogens may be lyophilised or frozen in aqueous solution for storage.

5.4 Immunisation

Few forensic laboratories are likely to have their own animal house for raising antisera, but contracting the work to a commercial organisation is both convenient and inexpensive.

Immunisation procedures are well described in the literature[7, 209] and will only be outlined here.

The choice of animal for immunisation depends on the amount of antiserum required. Rabbits yield 10–20 ml per bleed and are cheaper to purchase and house than sheep or goats which yield volumes of 150–300 ml per bleed. However, successive bleeds of high titre from a single rabbit can provide sufficient antiserum for many years of use. The response to immunisation varies widely and so several individual animals should be inoculated to ensure a reasonable chance of success.

Prior to inoculation, the drug-protein immunogen is blended with an adjuvant, a substance that enhances the immune response of the animal. Freund's complete adjuvant, which contains mineral oil, an emulsifier and killed Mycobacteria, is widely used for the primary inoculation. It may be used also for the subsequent booster inoculations although Freund's incomplete adjuvant which does not contain Mycobacteria is often used instead. The adjuvant functions as a sustained release medium for the immunogen while stimulating the reticulo-endothelial system and inducing an increased circulation of lymphocytes.

The aqueous immunogen (approximately 1 mg/ml) is blended with several volumes of adjuvant to form a stable, water-in-oil emulsion. A primary inoculation of about 100 μg of immunogen per rabbit (0.25–5 mg per sheep) is given either intramuscularly, intradermally or using a combination of sites. Other inoculation sites have been tried but there is no firm evidence that any particular site results in better antisera. Sites causing minimum discomfort to the animals should therefore be chosen.

The response to the primary inoculation is the formation of antibodies, mainly of the IgM class, whose concentration increases for several weeks. Booster inoculations are given at 1- to 6-monthly intervals after the primary inoculation, the doses of immunogen being about half that used for the primary inoculation. Antibodies of the IgG class are formed rapidly in response to a booster inoculation and reach their maximum serum concentration in about 10 days.

5.5 Collection and Storage of Antiserum

Blood is taken about 10–14 days after a booster inoculation. Several booster doses may be required to produce an antiserum of high titre and avidity.

The blood is allowed to clot and is then centrifuged to obtain the maximum amount of antiserum. When the antiserum has been tested (Sect. 3) and found to be satisfactory, it can be stored in various ways since IgG antibodies are relatively stable. The addition of 0.1% w/v sodium azide enables liquid antiserum to be stored at 4 °C without risk of bacterial degradation. Freezing at −20 °C or below is both effective and convenient as long as repeated freezing and thawing is avoided by aliquotting the antiserum prior to freezing. Lyophilisation of aliquots and storage at 4 °C or less is equally effective, though some antisera have been found to deteriorate when lyophilised[210].

A cloudy precipitate of denatured lipoprotein slowly forms in liquid antiserum but does not affect its performance in RIA. Ether extraction helps to avoid this but is not essential, nor is it necessary to isolate the IgG fraction of the antiserum by, for instance, ammonium sulphate precipitation. Affinity chromatography has been used to isolate the anti-drug IgG antibodies from antiserum[211–213] but, as far as RIA is concerned, there is little or nothing to be gained by this.

5.6 Monoclonal Antibodies

Monoclonal antibody production involves a combination of techniques from immunology and cell biology and yields limitless supplies of identical antibodies. Since the first report[214], the method has been described in detail[215–221] and monoclonal antibodies have been widely applied in various fields[222–225]. The original work was done with mice which is still the species of choise although monoclonal antibodies have since been raised using other species.

The principle of monoclonal antibody production is readily understood although the technical details are complex. When an animal is injected with an immunogen, antibodies are produced by cells in the spleen called B lymphocytes. Each B lymphocyte or its derived plasma cell produces a single type of antibody with a unique structure. Thus, if a single cell could be grown in culture ("cloned"), the resulting clone would yield monoclonal antibodies of identical structure and specificity. Unfortunately, normal antibody-producing cells cannot be cultured and so this simple approach to monoclonal antibody production is not possible. However, malignant tumours of the immune system called myelomas are known and these can not only be grown in culture but produce abnormal immunoglobulins called myeloma proteins. A myeloma is a clone formed from a single cell and so the myeloma proteins are affectively monoclonal antibodies. There is, however, no way of making a myeloma produce monoclonal antibodies in response to a particular immunogen.

The problem is surmounted by fusing the B lymphocytes from an immunised animal with myeloma cells. The hybrid melanoma or hybridoma cells that result combine the antibody-producing capability of the B lymphocytes with the immortality of the melanoma cells and so they produce antibodies and can be grown in culture.

Using a myeloma that produces no antibody of its own ensures that the monoclonal antibodies are derived solely from the B lymphocytes. Furthermore, if the myeloma cells are an enzyme deficient strain that dies unless cultured in a special medium, culturing the hybridoma cells in a different medium kills any unfused myeloma cells (unfused B lymphocytes also die) and simplifies the subsequent cloning and testing. An interesting observation[214, 215] is that the proportion of antibody-secreting hybridomas may be considerably higher than expected, indicating that certain B lymphocytes that manufacture but do not secrete antibodies form, on fusion with myeloma cells, hybridomas that secrete antibodies.

After cell fusion has been carried out, the surviving hybridomas are cultured separately and the supernatants of growing clones are tested for antibody activity. Clones secreting useful monoclonal antibodies are recloned before *in vitro* or *in vivo* production of antibodies in quantity.

The *in vitro* technique consists of culturing a suitable clone and harvesting the secreted monoclonal antibodies from the culture medium in which they may attain a concentration of 10–100 µg/ml. The *in vivo* technique involves injecting hybridoma cells into the peritoneal cavity of a mouse where they grow as a tumour, usually accompanied by ascites (up to about 15 ml) which is an accumulation of serous fluid in the peritoneal cavity. The ascites fluid contains a high concentration (1–10 mg/ml) of monoclonal antibody, and so a single mouse can yield over 100 mg of antibody. The ascites fluid is contaminated with normal mouse immunoglobulins and other serum proteins but, for RIA purpuses, this does not matter. By comparison, a polyclonal antiserum may contain 01.–1 mg/ml of numerous different antibodies against a particular immunogen. Culture supernatants and ascites fluids can be clarified by centrifugation and stored at 4 °C with sodium azide as a preservative. The effect of elevated temperatures, freeze/thaw cycles and pH changes on a variety of monoclonal antibodies has been documented[226].

Hybridoma cells can be stored in liquid nitrogen and recultured as required, and so monoclonal antibodies and be produced indefinitely and in any quantity. Although the initial investment is high compared with raising a polyclonal antiserum, the perpetuation of a successful clone is relatively inexpensive and so monoclonal antibody production is cost effective in the long term.

A disadvantage of monoclonal antibodies is that most of those produced after a successful cell fusion will be of low affinity. This is so with a polyclonal antiserum as well, the few high-affinity antibodies in the mixture determining the properties of the whole antiserum. A clone producing high-affinity monoclonal antibodies of the desired specificity can only be identified after a great deal of work, though recent advances in cell fusion techniques may improve the situation[224].

Monoclonal antibodies enable specific assays to be developed with no risk of exhausting the supply of antibody. The specificity of monoclonal antibodies against both proteins and small molecules does not, however, preclude unexpected cross-reactions[224]. Much of the research effort has been concerned with diagnostic assays of clinically important proteins with relatively little work so far on small-molecule assays[225]. A number of monoclonal antibodies against drugs and steroids have been raised[109, 227–238]. Most are comparable in performance to polyclonal antisera while some offer improved specificity.

6 Separation of the Bound and Free Fractions

Separation of the bound and free fractions is an important step in RIA and must be simple, rapid, reliable, economical and, above all, must not disturb the equilibrium. Various methods have been used and will be discussed briefly.

6.1 Widely-Used Separation Methods

6.1.1 Adsorption

Non-specific adsorption of the free fraction by a particulate material is simple, fast and cheap but the conditions must be optimised to avoid any significant disturbance of the equilibrium. The adsorbent is added to the equilibrium mixture and, after a specified interval for adsorption to occur, the mixture is centrifuged. The supernatant, i.e. the bound fraction, is pipetted or decanted into another vial for counting. Accurate timing between addition of the adsorbent and centrifugation is often a critical factor affecting the assay precision.

Powdered charcoal[239] sometimes but not always[240] coated with dextran or protein[241] is a commonly used adsorbent while alumina and various silicates have also been employed. Batchwise variations in properties, particularly with charcoal, can be a nuisance.

Other adsorbents such as hydroxyapatite[242] and zirconyl phosphate[243] adsorb the bound rather than the free fraction.

6.1.2 Fractional Precipitation

Fractional precipitation is widely used since it is not only simple, fast and cheap but highly reproducible. The addition of a salt, an organic solvent or polyethylene glycol (average molecular weight 6000–8000) to the equilibrium mixture removes water molecules from the system and disturbs the hydration shells around the antibody molecules, causing the antibodies to precipitate along with the bound antigen. Antibodies (γ-globulins) carry only a small charge at or near neutral pH and so are not highly solvated. As a result, relatively low concentrations of salts etc. can effect precipitation. In contrast, drug molecules are somewhat hydrophilic and remain in solution unless bound by the antibodies. Separation of the bound and free fractions is carried out by adding the precipitant, vortex mixing, centrifuging the precipitate, aspirating or decanting the supernatant, i.e. the free fraction, and counting the precipitate which is the bound fraction.

The most commonly used precipitants in drug RIA are ammonium sulphate, whose use for protein fractionation predates RIA, and polyethylene glycol[244]. An equal volume of saturated aqueous ammonium sulphate is added to the equilibrium mixture to give 50% saturation, at which concentration γ-globulins are readily precipitated. The optimum concentration of polyethylene glycol should be determined by experiment; 12.5% w/v in the equilibrium mixture is effective in RIAs for several protein hormones[244] while 17.5% is required in certain drug assays[71, 245]. Polyethylene glycol appears to be equally effective in buffer or aqueous solution.

Half saturated ammonium sulphate and 17.5% polyethylene glycol are both relatively viscous and so thorough vortex mixing after addition is essential. It is customary to allow 5–10 minutes after adding ammonium sulphate for complete precipitation to occur but, with polyethylene glycol, no delay is necessary.

The precipitates produced by ammonium sulphate are fairly bulky and more susceptible to disturbance than those produced by polyethylene glycol which are waxy and coherent. With careful aspiration, virtually 100% of an ammonium sulphate or polyethylene glycol supernatant can be removed. Washing the precipitates to remove trapped free fraction is not usually necessary since the slight gain in precision is offset by the extra work involved and the risk of disturbing the equilibrium.

In theory, ammonium sulphate and polyethylene glycol should give identical results but this does not always happen in practice and so it is worth trying both when developing an assay. The reasons for the difference are not known, though it has been shown that ammonium sulphate can disturb the antigen-antibody equilibrium[246].

The antigen-antibody reaction is slowed but not stopped by ammonium sulphate[246] or polyethylene glycol enabling "single reagent" assays to be developed in which the radioligand, antiserum and precipitating agent are added as a mixture to the sample[245, 246].

6.1.3 Second Antibody Precipitation

The second antibody technique[247] is widely used in RIA since it is universally applicable, simple, precise and has a minimal effect on the equilibrium. However, second antibodies are more expensive than ammonium sulphate or polyethylene glycol and more work is required to optimise the conditions than when fractional precipitation is used.

The principle of the method is lattice formation between the first, i.e. anti-drug, antibody and a second antibody raised against γ-globulins of the animal species that provided the first antibody. The second antibody must be course be raised in an unrelated species. For instance, immunising a donkey with rabbit γ-globulins produces a donkey anti-rabbit serum. The lattices that form are large enough to precipitate out of the solution and can be centrifuged, the first antibody in the lattice carrying with it the bound fraction. The supernatant containing the free fraction can be aspirated or decanted.

There is so litte γ-globulin in the dilute first antibody used in an assay that it is necessary to add normal serum or γ-globulin from the same species as the first antibody to ensure precipitation with the second antibody. Too little or too much second antibody impairs the formation of the precipitate and so the optimum proportions of first antibody, second antibody and serum or γ-globulin must be determined experimentally whenever a new bleed of second antibody is acquired.

Ammonium sulphate[248], dextran[248] or polyethylene glycol[249–252] can be included in the mixture since this can decrease the second antibody equilibration time and lower the amounts of second antibody and normal serum or γ-globulin that are required[250–252].

A second antibody separation can be carried out in several ways. The second antiserum can be added to the assay tubes after the antigen and first antiserum have reached equilibrium. A second equilibration is then required which may take several hours. This extra incubation can be avoided by adding the second antiserum and normal serum to the other reagents at the start of the assay. Alternatively, the first and second antisera can be premixed and the resulting precipitate used in the assay[253], as can the precipitate formed from second antiserum and normal serum[254]. Pre-precipitating the first antiserum adversely affects its titre and hence the sensitivity of the assay, while pre-precipitated γ-globulin from normal serum does not have this disadvantage and can be added in sufficient excess to reduce the second incubation to a few minutes[254].

Problems can arise in second antibody separations if the samples and standards differ in composition since various substances in serum can affect the rate of an antigen antibody reaction. An example is complement, a thermolabile component of serum which can be destroyed by heating or inactivated by the addition of a chelating agent such as EDTA to bind calcium and magnesium ions. Matrix effects can be avoided, however, by preparing standards in serum so that their composition resembles that of the samples.

Protein A, a cell-wall component of most strains of the bacterium *Staphylococcus aureus*, binds γ-globulin molecules specifically and can be used as an alternative to a second antibody. A suspension of intact bacteria[255] functions as a second antibody solid phase.

6.1.4 Solid Phase Separations

Attachment of a first or second antibody or protein A to a solid phase makes the mechanical separation of the bound and free fractions extremely simple, although some antibody activity may be lost if binding sites are blocked by the solid phase. The solid phase can be the wall of the assay tube or a separate solid material which may also be magnetic. An alternative solid phase is a polymerised antibody[256, 257] which does not need a separate solid support. There are numerous possible configurations of solid phase ranging from microparticulate powders to macrobeads, dipsticks, paddles, discs etc. Separation is effected by centrifuging, decanting, removing the solid support manually or, in the case of a magnetic solid phase, applying a magnetic field.

Solid phase methods are widely used in commercial assays but the work required for development and evaluation makes them less attractive for "in-house" drug RIAs unless the sample through-put is very high.

Antibodies can be attached to solid supports by physical, i.e. non-covalent, bonding or by chemical means[26]. The simplest physical method is adsorption from solution on to a plastic surface. The adsorption process is affected by such factors as temperature, pH, ionic strength, protein concentration and the nature of the surface. In practice, batch-to-batch reproducibility and desorption during an assay can present problems. Alternative non-covalent methods include entrapment of antibody in a gel or, for first antibody only, micro-encapsulation in a semi-permeable membrane.

Solid phases to which antibodies have been attached by covalent bonding include glass, polysaccharides such as agarose, cellulose and Sephadex, and plastics such as nylon, polyacrylamide, polyethylene, polypropylene, polystyrene, and polyvinyl chloride. The solid phase is activated[26] to introduce suitable electrophilic groups to which nucleophilic groups of the antibodies can be attached directly or via bifunctional reagents (Table 3). Polyacreolin microspheres, which have reactive aldehyde groups and require no activation step, have the additional advantage of being soluble in toluene-based liquid scintillants and so can be used in ^{3}H-labelled assays[258].

A novel solid phase technique involves the use of a reversibly soluble polymer consisting of first or second antibody bound to sodium alginate[259]. Adjustment of the pH or the addition of certain metal ions renders the polymer insoluble. The antigen-antibody reaction can thus be carried out in solution and the bound and free phases separated by precipitation once equilibrium has been attained.

Another novel solid phase consists of bismuth oxide and charcoal immobilised in starch spheres[260]. The free fraction is adsorbed by the charcoal and the spheres sediment in the assay tubes. The bismuth oxide attenuates the γ-rays emitted by the adsorbed radioligand and so the bound fraction in the supernatant can be counted without having to transfer it to a separate tube.

6.2 Less Widely-Used Separation Methods

6.2.1 Electrophoresis

This was used in the first RIA to be developed[1] but it is impractical and is now obsolete.

6.2.2 Gel Permeation Chromatography

Gel permeation chromatography is a useful technique for separating large and small molecules but it is cumbersome and little used as a separation method in RIA[261]. Membrane filtration is a more practical alternative and can be used in automated assay systems[262].

6.2.3 Ion Exchange

This is not a method of general applicability though it has been used in insulin[263] and vitamin B_{12}[264] assays.

6.2.4 Partition between Two Liquid Phases

Immunoassays using this type of separation have been described as "partition affinity ligand assays" (PALA)[265] and "ligand differentiation immunoassays" (LIDIA)[266, 267]. Such separations are speedy but are complicated to develop since modification of the radioligand or antibody may be necessary to confer preferential solubility on the free or bound fractions[266–268].

Direct solvent extraction into a liquid scintillant[74] or simple organic solvent[269] has been used to isolate the free fraction. When ^{125}I is used as the label, metal shielding around one of the separated phases allows the other to be counted without the need to decant into another vial. In a variant of this technique, enzyme hydrolysis of steroid glucuronide or sulphate in the free fraction releases steroid which is extracted into liquid scintillant[266]. Antibody-bound steroid glucoronide or sulphate resists hydrolysis and so it remains in the aqueous phase.

6.2.5 Dialysis

Dialysis has been used to separate bound and free testosterone[270]. Dissociation of the bound complex was minimised by a short dialysis time.

7 Quality Assurance

The quality assurance required in forensic drug RIA differs somewhat from that required in clinical drug monitoring. The most important parameter in forensic RIA is the distinction between positive and blank samples. In clinical analysis, quantitative accuracy is crucial but, in forensic RIA, the cross-reacting drugs and metabolites in a positive sample are identified and quantitated by alternative methods, the assay result indicating only the approximate overall level of cross-reacting material. Nevertheless, measures must be taken to ensure that a forensic drug RIA is performing reliably.

RIA quality assurance is complicated by the non-linearity and heteroscedasticity of the calibration curve. Commercially available software for RIA data processing generally includes some estimate of the error in extrapolated results, but uncritical reliance on this as an indicator of assay performance is unwise.

Quality assurance schemes described in the literature[21, 30, 271–277] are intended primarily for clinical RIA but the general principles are equally applicable to forensic RIA.

The major parameters of an assay that can be monitored for quality assurance purposes include:

(i) The total amount of radioligand added to the tubes. This is measured in the total tubes and indicates whether the radioligand has been diluted correctly.
(ii) The "zero binding", i.e. the % B_0^*/T for the zero standard. This shows whether the relative proportions of radioligand and antiserum are correct and whether the separation method is working properly.
(iii) The % B^*/T for the standard of highest concentration. This shows whether the overall assay response is stable. A significant change can indicate incorrectly-prepared or decomposing standards.
(iv) The shape and slope of the calibration curve. Significant variation will affect the precision and accuracy of the assay. For non-linear curves, the concentration corresponding to 50% B^*/T is a useful parameter.
(v) The agreement between replicate results. Poor replicates can indicate imprecision or error on the part of the operator.
(vi) The values given by blank and spiked samples. These should lie within reasonable limits (± 2 standards deviations).

In practice, the acceptable variation in the above parameters is established by repeating the assay at least ten times under the conditions in which it will be used, i.e. with different operators etc. Numerical values of some or all of the parameters are recorded in tabular or graphical form, as are the values for all subsequent assays. The data can be treated statistically but outliers or trends are usually obvious and indicate a problem with the assay. The most critical part of the curve in forensic RIA, the region around the detection limit, must of course be examined carefully.

It is impossible to define rigid guidelines that are valid for all assays; any deviation in a parameter must be evaluated with respect to the whole assay using common sense based on experience and erring on the side of caution.

8 Trouble-Shooting

Problems arise sooner or later in even the most robust and stable RIAs. Unless the cause can be identified easily, the simplest and often the quickest remedy is to discard all the reagents in use and start afresh. A number of problems and their likely causes are outlined below:

(i) Incorrect initial binding (% B_0^*/T). Antiserum and/or radioligand diluted incorrectly or wrong volumes added to tubes. Zero standard or diluent buffer contaminated with drug. Equilibrium not attained. Separation procedure not working correctly (check centrifuge as well as reagent).
(ii) Calibration curve too shallow or too steep. Concentrations of standards incorrect or wrong volumes added to tubes.
(iii) Poor replicates. Reagents not mixed thoroughly. Pipettes need cleaning and servicing. Viscous or clotted samples causing inaccurate pipetting. Carry-over from positive samples if pipette tip not changed between samples. Radiation counter faulty.

(iv) Wrong values from spiked samples. Samples prepared or diluted incorrectly. Wrong volumes added to assay tubes.

(v) Long-term drift in assay parameters. Radioligand, antiserum or standards unstable on storage.

Many problems can be avoided by attention to detail and the use of clearly-written method sheets. Pipettes should be cleaned and calibrated regularly. pH meters should be calibrated before use. Counters should be checked regularly and serviced annually. Centrifuges and other items of apparatus should be checked regularly. Buffers and other reagents held in stock should be dispensed in moderate amounts for daily use that can be discarded without undue waste should contamination be suspected. Antisera should be stored in aliquots to avoid contamination and unnecessary freezing and thawing. Frozen solutions should be thoroughly mixed before use. Standard solutions and spiked samples should be prepared and checked regularly. Standards and radioligands should be stored in suitable containers, e.g. silanised glass, in order to minimise adsorption on to the walls of the containers.

If Eppendorf-type pipettes are used, blood or other viscous samples are best transferred by reverse pipetting; the button on the pipette is depressed to the second stop, excess sample is drawn into the tip and the required amount is expelled by depressing the button to the first stop only. Alternatively, positive-displacement piston-type pipettes can be used and indeed are essential for pipetting small volumes of organic solvents or extracts accurately. Positive-displacement pipettes with disposable tips and pistons are now available and avoid any risk of carry-over from one sample to the next.

RIA should not be carried out in the same room as the analysis of drug seizures in order to avoid contamination by air-born powder. Analysts involved in both RIA and drug analysis should take care to avoid contamination of the assay tubes by drug residues on hands or clothing.

A problem that is more likely to occur with clinical rather than forensic specimens is contamination with an isotope used for nuclear diagnostic purposes[278]. This should be born in mind if inconsistent results are obtained repetitively.

9 Hazards

Regulations governing the possession, use and disposal of radioisotopes should be followed, but the hazard arising from the use of ^{125}I in RIA is negligible. The exposure rate at a distance of a few centimetres from the few kilobecquerels used in a typical assay is in the microsievert range. Even a manual radioiodination using 37 mBq (1 mCi) of ^{125}I carried out behind lead shielding to minimise whole-body exposure results in exposure to the fingers that is well below 1% of the dose limit (500 mSv/year) for individual organs and tissues specified in The Ionising Radiations Regulations 1985[279]. However, the minimal hazard should not be allowed to encourage complacency or the neglect of sensible precautions to minimise exposure and avoid accidents.

As in all forensic toxicology, hazards arise from the specimens analysed since hepatitis is prevalent amongst drug abusers and the number of AIDS carriers in increasing. No sample can be assumed to be free of risk and so all normal precautions should be taken at all times to avoid infection.

10 Published RIA Methods for Drug Analysis

This section lists drugs and other small molecules of forensic interest for which RIAs have been developed, though some references are to papers describing antiserum or radioligand preparation rather than complete RIA methods. Most drugs are grouped in therapeutic categories while some are grouped in chemical categories.

10.1 Analgesics and Narcotics

α-*l*-Acetylmethadol[147]
Alfentanil[280, 281]
Aminopyrine[282]
Anileridine and *N*-acetyl anileridine[283]
Antipyrine[284]
Aspirin[185, 285]
Buprenorphine[286]
Butorphanol[287]
Codeine[288–291]
Cyclazocine[148, 175]
Diphenoxylate[283]
Etorphine[292, 293]
Fentanyl[281, 294–298]
Hydrocodone[299]
Hydromorphone[288, 299, 300]
(-)-2-Hydroxy-*N*-cyclopropylmethylmorphinan[301]
Levorphanol[302, 303]
Meperidine[133, 199, 283] and normeperidine[283, 304]
Methadone[73, 140–142, 305–309]
Morphine (specific assays)[137, 138, 289, 310–314]
Morphine and related opiates (general assays)[56, 57, 135, 136, 193, 230, 245, 290, 305, 309, 315–337]
Pentazocine[145, 338–340]
Phenacetin[341]
Piminodine[283]
Sufentanil[280]

10.2 Antiasthmatics and Other Drugs Affecting the Respiratory System Including Bronchodilators, Bronchospasm Relaxants and Nasal Decongestants

d- and *l*-Ephedrine[151]
Formenterol[342]
d-Pseudoephedrine[343]
Salbutamol[344]
Theophylline[345, 346]

10.3 Antiobiotics and Antineoplastics

Actinomycin D[347, 348]
Adriamycin (Doxorubicin)[349–351]
Amikacin[352, 353]
1-β-*D*-Arabinofuranosylcytosine[354–356]
1-β-*D*-Arabinofuranosyluracil[357, 358]
Bleomycin[76, 159, 359–364]
Bruceantin[365]
Chloramphenicol[366–368]
Clindamycin[369]
Daunamycin[349]
9,3″-Diacetylmidecamycin[371]
2,3-Dihydro-1*H*-imidazo[1,2-*b*]pyrazole (IMPY)[372]
Etoposide[373]
Gentamycin[374–380]
Leucovorin[381]
Macromycin[382]
Methotrexate[78, 381, 383–394] and 7-hydroxymethotrexate[394]
5-Methyltetrahydrofolate[381]
Mitoxantrone[395]
Netilmicin[396]
Penicillin[397–403]
Peplomycin[404, 405]
L-Phenylalanine Mustard[406]
Rifampin[407]
Sisomicin[368, 408]
Sulphonamides[409, 410]
Tallysomycin S_{10b}[411]
Teniposide[373]
Tetracycline[412, 413]
Tobramycin[414]
Vinblastine[415–417], desacetylvinblastine[416] and 5′-noranhydrovine (Navelbine)[418]
Vincristine[415, 416]
Vindesine[416, 417]

10.4 Anticholinergic Drugs

Atropine[419, 420]
Carbachol[421]

10.5 Anticoagulants

Warfarin[143, 422, 423]

10.6 Anticonvulsants

Diphenylhydantoin (Phenytoin)[152, 424–432]
Phenobarbital (see Sect. 10.13.1)
Sulthiame[426]

10.7 Antidiabetic Drugs

Selected references to insulin and related protein hormones are included in this section since insulin is used occasionally for murder or suicide.

Buformin[433]
Glibenclamide[434–436]
Gliclazide[437]
Glipizide[438]
Gliquidone[436, 439]
Glisoxepide[440]
Insulin[1, 243, 253, 441–444]
  Endogenous insulin antibodies: free and bound insulin[445–451]
  Effect of haemolysis[451–455], hyperbilirubinaemia[455], hyperlipoproteinaemia[456] and uraemia[455] on insulin RIA
  Stability of insulin in serum[457, 458], plasma[458] and whole blood[459]
  Insulin in forensic cases[13, 460–473]
  Insulin in urine[474] and bile[475]
  C-Peptide[476–480] and proinsulin[481, 482] RIA

10.8 Antidiarrhoeal Drugs

Loperamide[483]

10.9 Antiemetic Drugs

Clebopride[484]

10.10 Antihistamines

Astemizole[485]
Brompheniramine[486]
Chlorpheniramine[486]
2-Methoxy-11-oxo-11-H-pyrido 2,1-*b* quinazoline-8-carboxylic acid[487]
Sodium cromoglycolate[488]
Terfenadine[206]
Triprolidine[489]

10.11 Anti-Inflammatory Drugs and Gout Suppressants

Antipyrine (see Sect. 10.1)
Aspirin (see Sect. 10.1)

Colchicine[195, 490, 491]
Indomethacin[492, 493]
Phenylbutazone[494]

10.12 Antimalarial Drugs

Chloroquine[495]
1-(1,3-Dichloro-6-trifluoromethyl-9-phenanthryl)-3-*N*,*N*-dibutylaminopropan-1-ol[149]
Quinine[496, 497]

10.13 Antipsychotic, Hypnotic and Sedative Drugs

10.13.1 Barbiturates

Barbiturates (general assays)[56, 57, 130, 309, 319, 498–506]
Barbital[507]
Cyclobarbital[282]
Phenobarbital[427, 432, 508–513]
R- and *S*-Secobarbital[144]

10.13.2 Benzodiazepines

Benzodiazepines (general assays)[72, 514–518]
Brotizolam[519]
Chlordiazepoxide[520–523]
Clonazepam[113, 432, 523–525]
Clozapine[526]
Diazepam[57, 71, 523, 527–531] and *N*-desmethyldiazepam[523]
Flunitrazepam[532]
Flurazepam[523, 533, 534]
Nitrazepam[523]
Oxazepam[523]
Prazepam[155]
Triazolam[535]

10.13.3 Phenothiazines

Phenothiazines (general assays)[52], [536]
Butaperazine[537, 538]
Chlorpromazine[177, 539–545], chlorpromazine sulphoxide[546] and 7-hydroxychlorpromazine[238]
Flupenthixol[547–549]
Fluphenazine[536, 549–552] and metabolites[536]
Perazine[537]
Perphenazine[553]
Prochlorperazine[537, 554]
Trifluoperazine[537, 549, 555–557], 7-hydroxytrifluoperazine[558], trifluoperazine $N^{4'}$-oxide[559] and trifluoperazine sulphoxide[560]
Trimeprazine[561]

10.13.4 Tricyclic Antidepressants

Tricyclic antidepressants (general assays)[52, 59, 114, 562–565]
Amitriptyline[60, 171, 543, 566–573]
Clomipramine[572–575]
Desmethylimipramine (desipramine)[543, 576, 577]
Doxepin[578, 579] and desmethyldoxepin[578, 579]
Imipramine[543, 577]
Nortriptyline[60, 171, 543, 566–571, 580–582]

10.13.5 Other Antipsychotic Drugs

Bupropion[583–585]
Haloperidol[186, 586–589] and reduced haloperidol[589]
Methaqualone[305, 590–593]
Nomifensine[594]
Pimozide[173, 595]
Sulpiride[596, 597]
Sultopride[596]

10.14 Antithyroid Drugs

Methimazole[598]
Propylthiouracil[58, 599]

10.15 Antitubercular Drugs

Isoniazid[600, 601] and isonicotinic acid[601]

10.16 Antiulcer Drugs

Carbenoxolone[602]
Pirenzapin[603]
Ranitidine[604, 605]

10.17 Antiviral Drugs

Antiviral drugs (review)[606]
Acyclovir[607, 608]
A134U[609]
BW A515U[610]
Cytosine arabinoside[611]
Ribavarin[612]

10.18 Anti-worm (Anthelmintic) Drugs

Oxfendazole[613]

10.19 Cardiovascular Drugs

10.19.1 Adrenergic Blocking Drugs

Acebutolol[614] and diacetolol[615]
Carazolol[616]
Propranolol[618–621]

10.19.2 Antiarrythmic Drugs

Aprindine[622]
Meobentine sulphate[623]
Procainamide[624, 625]
Quinidine[497, 626]

10.19.3 Antihypertensive Drugs

Captopril[627, 628]
Clonidine[629–633]
Debrisoquin[166]
4(5)-[2-(2,6-Dimethylphenyl)-ethyl]imidazole (MPV-295)[634]
Enalaprilat (active metabolite of enalapril)[634]
Guanabenz[636]
Hydralazine[637, 638]
Lisinopril[635]
Minoxidil[181]
Propranolol (see Sect. 10.19.1)
Reserpine[639]
Saralasin[640, 641]
Tiamenidine[198]

10.19.4 Cardiac Depressants

Alinidine[642]

10.19.5 Cardiac Glycosides and Aglycones

β-Acetyldigoxin[643, 644]
Acetylstrophanthidin[645, 646]
Deslanoside[647]
Digitonin[648]
Digitoxin[644, 649–654] and metabolites[654]
Digoxin (selected references only)
 Digoxin antibodies[200, 225, 227, 234, 655], radioligands[656, 657] and RIAs[658–667]
 Digoxin-like immunoreactive substances[668–680] and other factors[681–696] affecting results of digoxin RIAs
 Digoxin in forensic cases[697–720]
 Digoxin in neonates[721]
 Digoxin in erythrocytes[722]
 Digoxin in saliva[723]
 Digoxin in urine[724–726]
 Digoxin metabolites[654, 687, 727–731]

Gitaloxin[732, 733]
Gitoxin[732, 733]
Lanatoside C[647, 734, 735]
β-Methyldigoxin[647, 736–742]
Ouabain[743, 744]
Proscillaridin[745]

10.20 CNS Stimulants and Psychoactive Drugs

Amphetamine[51, 131, 172, 746–751] Note: (-)-deprenyl (selegiline)[752], methylamphetamine[753] and prenylamine[753] are metabolised in part to amphetamine
Caffeine[754]
Cannabinoids and metabolites[47, 48, 755–781]
 Lack of interference in cannabinoid RIA by endogenous compounds[782]
 Cannabinoids in forensic cases[783–789]
 Passive inhalation of cannabinoids[790–793]
 Cannabinoids in saliva[794]
Cocaine and benzoylecgonine[57, 96, 305, 331, 795–803]
2,5-Dimethoxy-4-methylamphetamine (DOM)[804–806]
3,4-Dimethoxyphenethylamine[804]
Lysergide (LSD)[807–818]
Methamphetamine[150, 174, 819–822]
Mescaline[804, 806, 823, 824]
Phencyclidine (PCP)[825–835]
Strychnine[836]

10.21 Diagnostic Aids

Metapyrone[837]

10.22 Diuretic Drugs

Bumetanide[838]
Piretanide[839]

10.23 Eisosanoids, Leukotrienes and Prostaglandins (Selected References)

Eicosanoids[840–842]
Leukotrienes[843–849]
Prostaglandins[849–855]
Sulprostone[856, 857]

10.24 Ergot Derivates

Dihydroergotamine[858–860]
Dihydroergotoxine[858, 861–864]

Ergometrine[858, 860]
Ergotamine[858]
Lisuride[865]
Lysergic Acid[866]
Lysergide (LSD) (see Sect. 10.20)
Methylergometrine[858]

10.25 Ganglionic Stimulant Drugs

Dimethylphenylpiperazinum (DMPP)[421]
Nicotine (see Sect. 10.33)

10.26 Herbicides and Insecticides

Aldrin[867]
S-Bivallethrin[146]
2,2bis-(*p*-chlorophenyl)1,1,1-trichloroethane (DDT)[868]
2,4-Dichlorophenoxyacetic acid (2,4-D)[869]
Dieldrin[867]
Malathion[868, 870]
Paraquat[182, 871–874]
Parathion[875, 876]

10.27 Immunomodulating and Immunosuppressive Drugs

Cyclosprin[877, 878]
erythro-9-(2-Hydroxy-3-nonyl)-hypoxanthine[879]

10.28 Muscle Relaxants

Decamethonium[421]
Ritodrine[880]
Succinylcholine[421]
d-Tubocurarine[881]

10.29 Narcotic Antagonists

(4)-2-Hydroxy-*N*-cyclopropylmethylmorphinan[882]
Nalmefene[883]
Naltrexone[883]
Naloxone[884, 885]

10.30 Renal Tubular Transport Inhibitors

Probenecid[886, 887]

10.31 Sympathomimetic Drugs

Adrenaline (*L*-epinephrine)[888]
Dopamine[888]
L-Dopa[888]
Noradrenaline (*L*-norepinephrine)[888]

10.32 Steroids (Selected References)

Sex determination by RIA of sex hormones[889–891]
Anabolic Steroids[892–896]
Corticosteroids[894, 897]
Betamethasone[898] and betamethasone 17-benzoate[899–901]
Budenoside[902]
Cyproterone acetate[903]
Dexamethasone[904–909]
Ethinyloestradiol[910–912]
Flunisolide[913]
Fluoxymesterone[914]
Hexoestrol[237]
Medroxyprogesterone[915–917]
Mestranol[910, 911, 916]
Methylprednisolone[918]
Metypyrone[919]
Norethindrone (norethisterone)[920–926]
Norgestrel[921, 927]
19-Nortestosterone[928–930]
Oxendolone[931]
3-Oxo desogestrel[932]
Prednisolone[933–939]
Prednisone[934–936, 940]
Triamcinolone acetonide[941, 942]
Trienbolone and trienbolone acetate[943]

10.33 Tobacco Alkaloids and Metabolites

N'-Acylnornicotine[944]
Cotinine[945–950]
Cotinine-N-oxide[945]
Nicotine[945, 946, 948, 951–954]
N'-nitrosonornicotine[944]
Nucleotide analogues of cotinine and nicotine[955]
γ-(3-Pyridyl)-γ-oxo-*N*-methylbutyramide[948, 956]

10.34 Toxins

Some protein toxins are included for their forensic interest

Abrin[957]
Aflatoxins[958–960]
Amanitins (amatoxins)[961–966]
Genistein[967]
Mycotoxins[968]
Ochratoxin A[969]
Paralytic shellfish poison[970]
Potato steroidal alkaloids[971]
Ricin[957]
Snake venoms[972]
Strychnine (see Sect. 10.20)

10.35 Vasodilators

Pencyclane[973]
Nicergoline[974]

10.36 Vitamins

Pantothenic acid[975, 976]
Vitamin A[977]
Vitamin D metabolites[978–981]
Vitamin K[982, 983]

11 References

1. Yalow RS, Berson SA (1959)Nature 184:1648
2. Ekins RP (1960) Clin Chim Acta 5:453
3. Skelley DS, Brown LP, Besch PK (1973) Clin Chem 19:146
4. Spector S (1973) Ann Rev Pharmacol 13:359
5. Marks V, Morris BA, Teale JD (1974) Br Med Bull 30:80
6. Butler VP (1975) J Immunol Methods 7:1
7. Landon J, Moffat AC (1976) Analyst 101:225
8. Butler VP (1977) J Pharm Exp Ther 29:103
9. Walker WHC (1977) Clin Chem 23:384
10. Paxton JW (1981) Meth Find Exptl Clin Pharmacol 3:105
11. Smith RN (1983) Anal Proc 20:417
12. Moffat AC (1975) Isolation and Identification of Drugs, Clarke EGC (ed) The Pharmaceutical Press, London vol 2 p 964
13. Smith RN (1980) Forensic Toxicology, Oliver JS (ed), Croom Helm, London, p 34
14. Smith RN (1984) Drug Determination in Therapeutic and Forensic Contexts, Reid E, Wilson ID (eds) Plenum Publishing Corporation, New York and London, p 261
15. Van Vunakis H, Langone JJ (eds) (1980) Methods in Enzymology 70
16. Odell WD, Daughaday WH (eds) (1971) Principles of Competitive Protein-Binding Assays. JB Lippincott Co., Philadelphia and Toronto
17. Mulé SJ, Sunshine I, Braude M, Willette RE (eds.) (1974) Immunoassays for Drugs Subject to Abuse. CRC Press, Cleveland
18. Pasternak CA (ed) (1975) Radioimmunoassay in Clinical Biochemistry. Heyden and Son Ltd., London

19. Ransom JP (1976) Practical Competitive Binding Assay Methods. The CV Mosby Co., St. Louis
20. Travis JC (1977) Fundamentals of RIA and other Ligand Assays. A Programmed Text. Radioassay Publishers, Anaheim
21. Chard T (1978) An Introduction to Radioimmunoassay and Related Techniques. North-Holland Publishing Co., Amsterdam
22. Voller A, Bartlett A, Bidwell D (eds) (1981) Immunoassays for the 80s. MTP Press Ltd., Lancaster
23. Langone JJ, Van Vunakis H (eds) (1982) Methods in Enzymology 84
24. Langone JJ, Van Vunakis H (eds) (1983) Methods in Enzymology 92
25. Butt WR (ed) (1984) Practical Immunoassay. The State of the Art. Marcel Dekker Inc., New York and Basel
26. Kricka LJ (1985) Ligand-Binder Assays. Labels and Analytical Strategies. Marcel Dekker Inc., New York and Basel
27. Ekins RP (1974) Br Med Bull 30:3
28. Feldman H, Rodbard D (1971) Principles of Competitive Protein-Binding Assays, Odell WD, Daughaday WH (eds), JB Lippincott Co., Philadelphia and Toronto, chap 7
29. Rawlins TGR, Yrjönen T (1978) Int Lab Nov./Dec.
30. Dudley RA, Edwards P, Ekins RP, Finney DJ, McKenzie IGM, Raab GM, Rodbard D, Rodgers RPL (1985) Clin Chem 31:1264
31. Scatchard G (1949) Ann NY Acad Sci 51:660
32. Berson SA, Yalow RS (1959) J Clin Invest 38:1996
33. Odell WD, Abraham G, Rand HR, Smerdloff RS, Fisher DA (1969) Acta Endocrinol Suppl 140:54
34. Keane PM, Walker WHC, Gauldie J, Abraham GE (1976) Clin Chem 22:70
35. Ekins RP, Newman GB, O'Riordan JLH (1968) Radioisotopes in Medicine: in Vitro Studies, Hayes RL, Goswitz FA, Murphy BEP (eds) U.S. Atomic Energy Commission, p 59
36. Yalow RS, Berson SA Radioisotopes in Medicine: in Vitro Studies, Hayes RL, Goswitz FA, Murphy BEP (1968) U.S. Atomic Energy Commission, p 7
37. Ekins RP (1969) Protein and Polypeptide Hormones, Margoulies M (ed) Excerpta Medica Foundation, Amsterdam, p 673
38. Ekins R, Newman B (1970) Acta Endocrinol Suppl 147:11
39. Ekins RP (1975) Radioimmunoassay in Clinical Biochemistry, Pasternak CA (ed) Heyden and Sons Ltd., London, p 3
40. Ekins R (1981) Immunoassays for the 80s (eds) Voller A, Bartlett A, Bidwell D, MTP Press Ltd., Lancaster, p 5
41. Ringle DA, Herndon BL (1972) J Immunol 109:174
42. Ryan JJ, Parker CW, Williams RC (1972) J Lab Clin Med 80:155
43. Weksler ME, Cherubin C, Kilcoyne M, Koppel G, Yoec M (1973) Clin Exp Immunol 13:613
44. Beranek JT, DeCato L, Adler FL (1976) Int Archs Allergy Appl Immunol 51:298
45. Beranek JT, DeCato L, Adler FL (1976) Int Archs Allergy Appl Immunol 51:402
46. Shapiro CM, Orlina AR, Unger PJ, Telfer M, Billings AA (1976) J Lab Clin Med 88:194
47. Teale JD, Forman EJ, King LJ, Piall EM, Marks V (1975) J Pharm Pharmacol 27:465
48. Williams PL, Moffat AC, King LJ (1978) J Chromatog 155:273
49. Beckett GJ, Hunter WM, Percy-Robb I (1978) Clin Chim Acta 88:257
50. Zettner A, Duly PE (1974) Clin Chem 20:5
51. Mason PA, Law B, Moffat AC (1983) J Immunoassay 4:83
52. Robinson K, Smith RN (1985) J Immunoassay 6:11
53. Stead AH, Moffat AC (1983) Hum Toxicol 2:437
54. Baselt RC (1982) Disposition of Toxic Drugs and Chemicals in Man, 2nd edn. Biomedical Publications, Davis, California
55. Colbert DL, Smith DS, Landon J, Sidki AM (1984) Clin Chem 30:1765
56. Usategui-Gomez M, Heveran JE, Cleeland R, McGhee B, Telischak Z, Awdziej T, Grunberg E (1975) Clin Chem 21:1378
57. Kaul B, Davidow B, Millian SJ (1977) J Analyt Toxicol 1:14
58. Cooper DS, Saxe VC, Maloof F, Ridgway EC (1981) J Clin Endocrinol Metab 52:204
59. Aherne GW, Piall EM, Marks V (1976) Br J Clin Pharmacol 3:561

60. Mould GP, Stout G, Aherne GW, Marks V (1978) Ann Clin Biochem 15:221
61. Sautebin L, Kindahl H, Kumlis L, Granstrom E (1985) Prostaglandins 30:435
62. Bayly RJ, Evans EA (1968) Storage and Stability of Compounds labelled with Radioisotopes, Review 7, The Radiochemical Centre, Amersham, U.K.
63. Evans EA (1976) Self-Decomposition of Radiochemicals, Review 16, The Radiochemical Centre, Amersham, U.K.
64. Krichevsky MI, MacLean CL (1971) Liquid Scintillation Counting, Dyer A (ed) Heyden and Son Ltd., London, chap 7
65. Assailly J, Bader C, Funck-Bretano J-L, Pavel D (1972) Liquid Scintillation Counting, Crook MA, Johnson P, Scales B (eds) Heyden and Son Ltd., London, 2 chap 19
66. Crook MA, Johnson P (eds) Liquid Scintillation Counting,Heyden and Son Ltd., London, 3 chaps 5–9
67. Dyer A (1974) An Introduction to Liquid Scintillation Counting, Heyden and Son Ltd., London
68. Peng CT (1977) Sample Preparation in Liquid Scintillation Counting, Review 17, The Radiochemical Centre, Amersham, UK
69. Wagstaff HS, Ware AR (1977) Liquid Scintillation Counting, Crook MA, Johnson P (eds), Heyden and Son Ltd., London, 4, chap 8
70. Bray GA (1960) Anal Biochem 1:279
71. Rutterford MG, Smith RN (1980) J Pharm Pharmacol 32:449
72. Robinson K, Rutterford MG, Smith RN (1980) J Pharm Pharmacol 32:773
73. Robinson K, Smith RN (1983) J Pharm Pharmacol 35:566
74. Jowett TP, Slater JDH, Piyasena RD, Ekins RP (1973) Clin Sci Mol Med 45:607
75. Purchas RW, Zinn SA, Tucker HA (1985) Anal Biochem 149:399
76. Elson MK, Oken MM, Shafer RB (1977) J Nucl Med 18:296
77. Harwig JF, Oflaz G, O'Brien TJ, Nakamura RM, Wolf W (1983) Int J Nucl Med Biol 10:263
78. Paxton JW, Rowell FJ, Cree GM (1978) Clin Chem 24:1534
79. Freedlender AE (1969) Protein and Polypeptide Hormones, Margoulies M (ed) Excerpta Medica Foundation, Amsterdam, pp 351 and 611
80. Seevers RH, Counsell RE (1982) Chem Rev 82:575
81. Dewanjee MK, Rao SA (1983) Radioisotopes for Medical Applications, Rayuda GVS (ed) CRC Press, Cleveland, 2, chap 1
82. Hunter WM, Greenwood FC (1962) Nature 194:495
83. Greenwood FC, Hunter WM, Glover JS (1963) Biochem J 89:114
84. Hughes WL (1957) Ann NY Acad Sci 70:3
85. Hunter R (1970) Proc Soc Exp Biol Med 133:989
86. Tantchou JK, Slaunwhite WR (1979) Prep Biochem 9:379
87. Hanson RN (1985) Int J Nucl Med Biol 12:315
88. Smith RN Unpublished Results
89. Fraker PJ, Speck JC (1978) Biochem Biophys Res Comm 80:849
90. Reay P (1982) Ann Clin Biochem 19:129
91. Boothe TE, Finn RD, Vora MM, Emran AM, Kathari PJ (1984) Int J Appl Radiat Isot 35:1138
92. Boothe TE, Finn RD, Vora MM, Emran AM, Kathari PJ, Kabalka K (1985) J Labell Comp Radiopharm 22:1109
93. Meyers J, Krohn K, DeNardo G (1975) J Nucl Med 16:835
94. Eckelman WC, Kubotta H, Siegel BA, Komai T, Rzeszotarski WJ, Reba RC (1976) J Nucl Med 17:385
95. McFarlane AS (1958) Nature 182:53
96. Robinson K, Smith RN (1984) J Pharm Pharmacol 36:157
97. England BG, Niswender DG, Midgley AR (1974) J Clin Endocrinol Metab 38:42
98. Hoffmann K, Samajeewa P, Smith ER, Kellie AE (1975) J Steroid Biochem 6:91
99. Allen RM, Redshaw MR (1978) Steroids 32:467
100. Corrie JET, Ratcliffe WA, Macpherson JS (1981) Steroids 38:709
101. Nordblom GD, Webb R, Counsell RE, England BG (1981) Steroids 38:161
102. Berg D, Thaler F, Kuss E (1982) Acta Endocrinol 100:154
103. Corrie JET, Ratcliffe WA, Macpherson JS (1982) J Immunol Meth 51:159

104. Fránek M, Bursa J, Hruška K (1983) J Steroid Biochem 19:1371
105. Corrie JET, Hunter WM, Macpherson JS (1980) J Endocrinol 87:8P
106. Corrie JET, Hunter WM, Macpherson JS (1981) Clin Chem 27:594
107. Chaudhri R, Coulson WF (1982) J Steroid Biochem 16:87
108. Tiefenauer LX, Andres RY (1984) J Immunol Meth. 74:293
109. Brochu M, Veilleux R, Lorrain V, Belanger A (1984) J Steroid Biochem 21:405
110. Puizillout JJ, Delaage MA (1981) J Pharmacol Exp Ther 217:791
111. Cailla HL, Racine-Weisbuch MS, Delaage MA (1973) Anal Biochem 56:394
112. Bolton AE, Hunter WM (1973) Biochem J 133:529
113. Dixon R, Crews T (1977) Res Commun Chem Pathol Pharmacol 18:477
114. Mason PA, Rowan KM, Law B, Moffat AC, Kilner EA, King LA (1984) Analyst 109:1213
115. Word FT, Wu MM, Gerhart JC (1975) Anal Biochem 69:339
116. Reiner L, Keston AS, Green M (1943) Science 96:362
117. Carraway KL (1975) Biochim Biophys Acta 415:379
118. Schlager SI (1980) Methods in Enzymology 70:247
119. Assoian RK, Blix PM, Rubenstein AH, Tager HS (1980) Anal Biochem 103:70
120. Gabel C, Shapiro BM (1978) Anal Biochem 86:396
121. Rand-Weaver M, Abuknesha RA, Price RG (1985) FEBS Lett 182:185
122. Samuel D, Amlot PL, Abuknesha RA (1985) J Immunol Meth 81:123
123. Ruliffson WS, Lang HM, Hummel JP (1953) J Biol Chem 201:839
124. Randerath K (1981) Anal Biochem 115:391
125. Heindel ND, Van Dort M (1985) J Org Chem 50:1988
126. Tae H, Inhae J (1982) Anal Biochem 121:286
127. Mitchell GA, Smutny PV, Chambliss KW, Levine BD (1974) Immunochem. 11:611
128. Morris BJ (1976) Clin Chim Acta 73:213
129. Bürgisser E (1984) J Receptor Res 4:357
130. Mason PA, Law B, Pocock K, Moffat AC (1982) Analyst 107:629
131. Mason PA, Bal TS, Law B, Moffat AC (1983) Analyst 108:603
132. Law B, Mason PA, King LJ, Moffat AC (1984) J Anal Toxicol 8:14
133. Robertson M (1982) New Scientist 696
134. Landsteiner K (1945) The Specificity of Serological Reactions, Harvard University Press, Cambridge, Massachusetts
135. Spector S (1971) J Pharmacol Exp Ther 178:253
136. Wainer BH, Fitch FW, Fried J, Rothberg RM (1973) J. Immunol. 110:667
137. Gross SJ, Grant JD, Wong SR, Schuster R, Lomax P, Campbell DH (1974) Immunochem. 11:453
138. Morris BA, Robinson JD, Piall E, Aherne GW, Marks V (1974) J Endocrinol 64:6
139. Lindner HR, Perel E, Friedlander A, Zeitlin A (1972) Steroids 19:357
140. Bartos F, Olsen GD, Leger RN, Bartos D (1977) Res Commun Chem Pathol Pharmacol 16:131
141. Bartos F, Bartos D, Olsen GD, Anderson B, Daves GD (1978) Res Commun Chem Pathol Pharmacol 20:157
142. McGilliard KL, Wilson JE, Olsen GD, Bartos F (1979) Proc West Pharmacol Soc 22:463
143. Cook CE, Ballentine NH, Seltzman TB, Tallent CR (1979) J Pharmacol Exp Ther 210:391
144. Cook CE, Myers MA, Tallent CR, Seltzman T, Jeffcoat AR (1979) Fed Proc 38:742
145. Nambara T, Tanaka T, Numazawa M (1979) J Pharm Dyn 2:151
146. Wing KD, Hammock BD (1979) Experientia 35:1619
147. McGilliard KL, Olsen GD (1980) J Pharmacol Exp Ther 215:205
148. Maeda M, Tsuji A (1981) J Pharm Dyn 4:167
149. Cook CE, Seltzman TP, Tallent CR, Wooten JD (1982) J Pharmacol Exp Ther 220:568
150. Niwaguchi T, Kanda Y, Kishi T, Inoue T (1982) J Forensic Sci 27:592
151. Midha KK, Hubbard JW, Cooper JK, MacKonka C (1983) J Pharm Sci 72:736
152. Paxton JW, Rowell EJ, Ratcliffe JG (1976) J Immunol Meth 10:317
153. Beckett GJ, Hunter WM, Percy-Robb IW (1978) Clin Chim Acta 88:257
154. Erlanger BF (1980) Methods in Enzymology 70:85
155. Köhler-Schmidt H, Bohn G (1983) Forensic Sci Int 22:243
156. Erlanger BF (1973) Pharm Rev 25:271

157. Kennedy JH, Kricka LJ, Wilding P (1976) Clin Chim Acta 70:1
158. Riceberg LJ, Van Vunakis H, Levine L (1974) Anal Biochem 60:551
159. Seki T, Muraoka Y, Takahashi K, Horinishi H, Umezawa H (1985) J Antibiotics 38:1251
160. Wold F (1972) Methods in Enzymology 25:623
161. Peters K, Richards FM (1977) Ann Rev Biochem 46:523
162. Avreamus S, Ternynck T, Guesdon J-L (1978) Scand J Immunol 8 (Suppl. 7):7
163. Ishikawa E (1980) Immunoassay, Suppl. 1, University of Sheffield Biomedical Information Service, Sheffield, UK
164. Cammisuli S (1981) Immunological Methods, Lefkovits I, Pernis B. (eds), Academic Press, New York, chap. 7
165. Fujiwara K, Yasuno M, Kitigawa T (1981) J Immunol Meth 45:195
166. Dixon R, Fahrenholtz KE, Burger W, Perry C (1977) Res Commun Chem Pathol Pharmacol 16:121
167. Ternynck T, Avreamus S (1976) Ann Immunol (Paris) 127C:197
168. Wilson DV, Devey M (1973) Int Arch Allergy 44:77
169. Shafer J, Baronowsky P, Laursen R, Finn F, Westheimer FH (1966) J Biol Chem 241:421
170. Surinov BP, Manoylov SE (1966) Biokhimiia (Moskva) 31:337
171. Kamel RS, Landon J, Smith DS (1979) Clin Chem 25:1997
172. Cheng, LT, Kim SY, Chung A, Castro A (1973) FEBS Lett 36:339
173. Michiels LJM, Heykants JJP, Knaeps AG, Janssen PAJ (1975) Life Sci 16:937
174. Inayama S, Tokunaga Y, Hosoya E, Nakadate T, Niwaguchi T, Aoki K, Saito S (1977) Chem Pharm Bull 25:838
175. Tsuji A, Maeda M, Kosage H (1980) J Pharmacobio-Dynam 3:S22
176. Cook CE, Kepler JA, Christensen HD (1973) Chem Pathol Pharmacol 5:767
177. Hubbard JW, Midha KK, McGilveray J, Cooper JK (1978) J Pharm Sci 67:1563
178. Creech HJ, Franks WR (1937) Amer J Cancer 30:555
179. Creech HJ, Franks WR (1938) J Amer Chem Soc 60:127
180. Morley JS, Capper SJ, Miles J (1984) Neuropeptides 4:477
181. Royer ME, Ko H, Gilbertson TJ, McCall JM, Johnston KT, Stryd R (1977) J Pharm Sci 66:1266
182. Fatori D, Hunter WM (1980) Clin Chim Acta 100:81
183. Wainer BH, Wung, WE, Hill JH, Fitch FW, Fried J, Rothberg RM (1976) J Pharmacol. Exp. Ther 197:734
184. Weiner LM, Rosenblatt M, Howes H (1963) J Immunol 90:788
185. Butler GC, Harington CR, Yeuill ME (1940) Biochem J 34:838
186. Wurzburger RJ, Miller RL, Marcum EA, Colburn WA, Spector S (1981) J Pharmacol Exp Ther 217:757
187. Chapman RS, Ratcliffe JG (1982) Clin Chim Acta 118:129
188. Newby A, Sala GB (1982) Biochem J 208:603
189. Gray GR (1978) Methods in Enzymology 50:155
190. Zopf D, Smith DF, Drzeniek Z, Tsai CM, Ginsburg V (1978) Methods in Enzymology 50:171
191. Zopf D, Tsai C-M, Ginsburg V (1978) Methods in Enzymology 50:163
192. Smith DF, Zopf D, Ginsburg V (1978) Methods in Enzymology 50:169
193. Spector S, Parker CW (1970) Science 168:1347
194. Africa B, Haber A (1971) Immunochem 8:479
195. Pontikis R, Scherrmann JM, Nam N-H, Boudet L, Pichat L (1980) J Immunoassay 1:449
196. Ungar-Waron R, Sela M (1966) Biochim Biophys Acta 124:147
197. Cordoba F, Gonzalez L, Rivera P (1966) Biochim Biophys Acta 127:151
198. Eckert HG, Baudner S, Weimer KE, Wissmann H (1981) Arneim-Forsch/Drug Res 31:419
199. Catlin DH, Schaeffer JC, Fischer JF (1975) Res Commun Chem Pathol Pharmacol 11:245
200. Thong, B, Soldin SJ, Lingwood CA (1985) Clin Chem 31:1625
201. de Weck AL, Eisen HN (1960) J Exp Med 112:1227
202. Weinstein A, Lindner HR, Friedlander A, Bauminger S (1972) Steroids 20:789
203. Erlanger BF, Borek F, Beiser SM, Liebermann S (1957) J Biol Chem 228:713
204. Kamel RS, Landon J, Smith DS (1978) Clin Chim Acta 84:403
205. Abraham GE, Swerdloff R, Tulchinsky D, Odell WD (1971) J Clin Endocrinol Metab 32:619

206. Cook CE, Williams DL, Myers M, Tallent CR, Leeson GA, Okerholm RA, Wright GJ (1980) J Pharm Sci 69:1419
207. Yoshizawa I, Kameyama M (1985) J Chromatog 338:404
208. Hodgkinson AJ, Sidki AM, Landon J, Rowell FJ, Mahmod SM (1985) Ann Clin Biochem 22:519
209. Hurn BAL, Chantler SM (1980) Methods in Enzymology 70:104
210. Al-Tamer YY, Taylor A, Marks V (1977) Clin Chem 23:2175
211. Spratt JL, Jones SB (1976) Life Sci 18:1013
212. Walker MC, Simon EJ (1977) J Pharmacol Exp Ther 202:360
213. Wahid FA, Isom GE (1980) Res Commun Substance Abuse 1:451
214. Köhler G, Milstein C (1975) 256:495
215. Milstein C (1980) Scientific American 243:56
216. Kennett RH, McKean TJ, Bechtol KB (eds) (1980) Monoclonal Antibodies. Hybridomas: a new Dimension in Biological Analyses. Plenum Press, London and New York
217. Hurn BAL, Chantler SM (1980) Methods in Enzymology 70:135
218. Albertini A, Ekins R (eds.) (1981) Monoclonal Antibodies and Developments in Immunoassay. Elsevier, Amsterdam, New York and Oxford
219. Hämmerling GJ, Hämmerling U, Kearney JF (eds) (1981) Monoclonal Antibodies and T-Cell Hybridomas. Elsevier, Amsterdam, New York and Oxford
220. Köhler G (1981) Immunological Methods, Lefkovits I, Pernis B (eds), Academic Press, New York and London, vol 2, p 285
221. Langone JJ, Van Vunakis H (eds) (1983) Methods in Enzymology, Academic Press, New York and London, vol 92
222. Edwards PAW (1981) Biochem J 200:1
223. King JS (1985) Clin Chem 31:1420
224. Scott MG (1985) Trends in Biotechnology 3:170
225. Zalcberg JR (1985) Pharmacol Ther 28:273
226. Underwood PA, Bean PA (1985) J Immunol Meth 80:189
227. Bang BE, Hurme M, Juntunen K, Mäkelä O (1981) Scand J Clin Lab Invest 41:75
228. Kohen F, Lichter S, Eshar Z, Linder HR (1982) Steroids 39:453
229. Mudgett-Hunter M, Margoulies MN, Ju A, Haber E (1982) J Immunol 129:1165
230. Glasel JA, Bradbury WM, Venn RF (1983) Molecular Immunol 20:1419
231. Place JD, Thompson SG, Clements HM, Ott RA, Jensen FC (1983) Antimicrob Agents Chemother 24:246
232. Zalcberg JR, Healey K, Hurrell J, McKenzie IFC (1983) Int J Immunopharmac 5:397
233. Al-Dujaili EAS, Hubbard AL, van Heyningen V, Edwards CRW (1984) J Steroid Biochem 20:849
234. Collignon A, Edelman L, Scherrmann JM (1984) Int J Nucl Med Biol 11:97
235. Fantyl VE, Wang DY (1984) J Endocrinol 100:367
236. Kato Y, Paterson A, Langone JJ (1984) J Immunol Meth 67:321
237. Carter APD, Cottingham JD, Dixon SN, Heitzmann RJ (1985) J Vet Pharmacol Ther 8:362
238. Yeung PKF, McKay G, Ramshaw IA, Hubbard JW, Midha KK (1985) J Pharmacol Exp Ther 233:816
239. Herbert V, Lau KS, Gottlieb L, Bleicher SJ (1965) J Clin Endocrinol Metab 25:1375
240. Ekins RP (1969) Protein and Polypeptide Hormones, Margoulies M (ed.), Excerpta Medica Foundation, Amsterdam, p 633
241. Boxen I, Tevaarwerk GJM (1982) J Immunoassay 3:53
242. Trafford DJH, Ward PR, Foo AY, Makin HLJ (1976) Steroids 27:405
243. Coffey JW, Nagy CF, Lenusky R, Hansen HJ (1974) Biochem Med 9:54
244. Desbuquois B, Aurbach GD (1971) J Clin Endocrinol Metab 33:732
245. Smith RN (1981) J Immunoassay 2:75
246. Seppälä IJT (1975) J Immunol Meth 9:135
247. Utiger RD, Parker ML, Daughaday WH (1962) J Clin. Invest. 41:254
248. Martin MJ, Landon J (1975) Radioimmunoassay in Clinical Biochemistry, Pasternak CA (ed), Heyden and Son Ltd., London, p 269
249. Catty D, King TD (1974) Immunochem 11:615

250. Remeš V, Slaninová J, Škopková J, Barth T, Franěk F (1978) Radiochem Radioanal Lett 34:253
251. Ansari AA, Bahuguna LM, Malling HV (1979) J Immunol Meth 26:203
252. Henley R, Kisgyorgy D (1980) Med Lab Sci 37:367
253. Hales CN, Randle PJ (1963) Biochem J 88:137
254. Midgley AR, Hepburn MR (1980) Methods in Enzymology 70:266
255. Brunda MJ, Minden P, Sharpton TR, McClatchy JK, Farr RS (1977) J Immunol 119:193
256. Donini S, Donini P (1969) Acta Endocrinol Suppl 142:257
257. Chesworth JM (1977) Anal Biochem 80:31
258. Pines M, Margel S (1986) J Immunoassay 7:97
259. Marshall DL (1982) Anal Lett 15:1457
260. Eriksson H, Mattiason B, Thorell JI (1981) J Immunol Meth 42:105
261. Haber E, Page LB, Richards FF (1965) Anal Biochem. 12:163
262. Goldie DJ, West PM, Ismail AAA, Poole AC, Thomas MJ (1983) Immunoassays for Clinical Chemistry, Hunter WM, Corrie JET (eds), Churchill Livingstone, Edinburgh, p 226
263. Meade RL, Klitgaard HM (1962) J Nucl Med 3:407
264. Frenkel EP, Keller S, McCall MS (1966) J Lab Clin Med 68:510
265. Mattiasson B (1980) J Immunol Meth. 35:137
266. Barnard GJ, Matson CM, Makawiti DW, Kilpatrick MJ, Collins WP (1983) Immunoassays for Clinical Chemistry, Hunter WM, Corrie JET (eds), Churchill Livingstone, Edinburgh, p 231
267. Makawiti DW, Barnard GJR, Matson CM, Collins WP (1983) J Steroid Biochem 18:619
268. Mattiasson B, Ling TGI (1980) J Immunol Meth 38:217
269. Cais M, Shimoni M (1981) Ann Clin Biochem 18:317
270. Rommerts FFG, Clotscher WF, Van der Molen HJ (1977) Anal Biochem 82:503
271. Rodbard D (1974) Clin Chem 20:1255
272. Woodcock, BG, Rodgers M, Gibbs AC (1975) Radioimmunoassay in Clinical Biochemistry, Pasternak, CA (ed), Heyden and Son Ltd, London, p 131
273. Woo J, Cannon, DC (1976) Amer J Clin Pathol 66:854
274. Cerceo E (1976) Lab Practice 25:685
275. Kemp KW, Nix ABJ, Wilson DW, Griffiths K (1978) J. Endocrinol 76:203
276. Hunter WM, McKenzie I, Bacon RRA (1981) Immunoassays for the 80s, Voller A, Bartlett A, Bidwell D (eds), MTP Press Ltd, Lancaster, p 155
277. Woodhead JS, Kemp, HA, Nix ABJ, Rowlands RJ, Kemp HW, Wilson DW, Griffiths K (1981) Immunoassays for the 80s, Voller A, Bartlett A, Bidwell D (eds), MTP Press Ltd, Lancaster, p 169
278. Solomon HM, Reich SD (1970) Lancet 1038
279. The Ionising Radiations Regulations 1985, HMSO, London
280. Michiels M, Hendriks R, Heykants J (1982) J. Pharm Pharmacol 35:86
281. Schüttler J, White PF (1984) Anesthesiol 61:315
282. Terazawa K, Takatori T (1982) J Forensic Sci 27:844
283. Van Vunakis H, Freeman DS, Gjika HB (1975) Res Commun Chem Pathol Pharmacol 12:379
284. Chang RL, Wood AW, Dixon WR, Conney AH, Anderson KE, Eiseman J, Alvares AP (1976) Clin Pharmacol Ther 20:219
285. Weiner LM, Rosenblatt M, Howes HA (1963) J Immunol 90:788
286. Bartlett AJ, Lloyd-Jones JG, Rance MJ, Flockhart IR, Dockray GJ, Bennett MRD, Moore RA (1980) Eur J Clin Pharmacol 18:339
287. Pittman KA, Smyth RD, Mayol RF (1980) J Pharm Sci 69:160
288. Wainer BH, Fitch FW, Fried J, Rothberg RM (1974) Clin Immunol Immunopathol 3:155
289. Findlay JWA, Butz RF, Welch RM (1977) Res Commun Chem Pathol Pharmacol 17:595
290. Gintzler AR, Mohacsi E, Spector S (1976) Eur J Pharmacol 38:149
291. Findlay JWA, Butz RF, Welch RM (1976) Life Sci 19:389
292. Robinson JD, Morris BA, Marks V (1975) Res Commun Chem Pathol Pharmacol 10:1
293. Woods WE, Weckman T, Wood T, Chang S-L, Blake JW, Tobin T (1986) Res Commun Chem Pathol Pharmacol 52:237

294. Henderson G, Frincke J, Leung CY, Torten M, Benjamini E (1974) Proc West Pharmacol Soc 17:64
295. Henderson GL, Frincke J, Leung CY, Torten M, Benjamini E (1975) J Pharmacol Exp Ther 192:489
296. Schleimer R, Benjamini E, Eisele J, Henderson G (1976) Proc West Pharmacol Soc 19:237
297. Michiels M, Hendriks R, Heykants J (1977) Eur J Clin Pharmacol 12:153
298. Woods WE, Tai HH, Tai C, Weckman T, Wood T, Barios H, Blake JW, Tobin T (1986) Amer J Vet Res 47:2180
299. Honigberg IL, Stewart JT (1980) J Pharm Sci 69:1171
300. Honigberg IL, Stewart JT, Brown WJ, Jun HW, Needham TE, Vallner JJ (1977) J Anal Toxicol 1:70
301. Dixon R, Young R, Mohacsi E, Holazo A, Malarek D, Boxenbaum H, Moore J, Parsonnet M, Kaplan S (1978) Clin Pharmacol Ther 24:622
302. Dixon R, Crews T, Mohacsi E, Inturrisi C, Foley K (1980) Res Commun Chem Pathol Pharmacol 29:535
303. Dixon R, Crews T, Mohacsi E, Inturrisi C, Foley K (1981) Res Commun Chem Pathol Pharmacol 32:545
304. Freeman DS, Gjika HB, Van Vunakis H (1977) J Pharmacol Exp Ther 203:203
305. Cleeland R, Christenson J, Usategui-Gomez M, Heveran J, Davis R, Grunberg E (1976) Clin Chem 22:712
306. Manning T, Bidanset JH, Cohen S, Lukashi L (1976) J Forensic Sci 21:112
307. Roerig DL, Wang RIH, Mueller MM, Lewand DL, Adams SM (1976) Clin Chem 22:1915
308. Ling GSF, Umans JG, Inturrisi CE (1981) J Pharmacol Exp Ther 217:147
309. Ziminski, KR, Wemyss CT, Bidanset JH, Manning TJ, Lukash L (1984) J Forensic Sci 29:903
310. Grabinsky, PY, Kaido RF, Walsh TD, Foley KM, Houde RW (1983) J Pharm Sci 72:27
311. Moore RA, Baldwin D, Allen MC, Watson PJQ, Bullingham RES, McQuay HJ (1984) Ann Clin Biochem 21:318
312. Sandouk P, Scherrmann JM, Bourdon R (1984) J Pharmacol Meth 11:227
313. Edwards DJ, Popovski Z, Baumann TJ, Bivins BA (1986) Clin Chem 32:157
314. Loeber JG, Elvers LH, Somers HHJ (1986) Clin Chem 32:1790
315. Wainer BH, Fitch FW, Rothberg RM, Fried J (1972) Science 176:1143
316. Wainer BH, Fitch FW, Rothberg RM, Fried J (1972) Science 178:647
317. Van Vunakis H, Wasserman E, Levine L (1972) J Pharmacol Exp Ther 180:514
318. Catlin D, Cleeland R, Grunberg E (1973) Clin Chem 19:216
319. Spector S, Berkowitz B, Flynn EJ, Peskar B (1973) Pharmacol Rev 25:281
320. Koida M, Takahashi M, Kaneto H (1974) Japan J Pharmacol 24:707
321. Koida M, Takahashi M. Muraoka S, Kaneto H (1974) Japan J Pharmacol 24:165
322. Capelle N, Fournier E, Kuntz MO, Treich I (1974) J Europ Toxicol 7:147
323. Davis R, Feldhaus J, Heveran J, Wicks R, Peckham M (1975) Clin Chem 21:1498
324. Jain NC, Sneath TC, Leung WJ, Budd RD (1977) J Pharm Sci 66:66
325. Möller MR, Tausch D, Biro G (1977) Z Rechtsmed 79:103
326. Baumgartner AM, Jones PF, Baumgartner WA, Black CT (1979) J Nucl Med 20:748
327. Hahn EF, Fishman JH, Fishman J (1979) J Pharm Pharmacol 31:250
328. Nelson PE, Fletcher SM, Moffat AC (1980) J Forensic Sci Soc 20:195
329. Findlay JWA, Butz RF, Jones EC (1981) Clin Chem 27:1524
330. Mason PA, Law B, Ardrey RE (1981) J Label Comp Radiopharm 18:1497
331. Valente D, Cassini M, Pigliapochi M, Vansetti G (1981) Clin Chem 27:1952
332. Spector S (1982) Methods in Enzymology 84:551
333. Hsu A-F, Brower D, Etskovitz RB, Chen PK, Bills DB (1983) Phytochem 22:1665
334. Püschel K, Thomasch P, Arnold W (1983) Forensic Sci Int 21:181
335. Svenneby G, Wedege E, Karlsen KL (1983) Forensic Sci Int 21:223
336. Aherne GW, Littleton P (1985) Lancet 210
337. Matejczyk RF, Kosinski JA (1985) J Forensic Sci 30:677
338. Williams TA, Pittman KA (1973) Fed Proc 32:719
339. Williams TA, Pittman KA (1974) Res Commun Chem Pathol Pharmacol 7:119

340. Peterson JE, Graham M, Banks WF, Benziger D, Rowe EA, Clemans S, Edelson J (1979) J Pharm Sci 68:626
341. Findlay JWA, De Angelis RL, Butz RF, Sailstad JB, Welch RM (1979) J Pharmacol Exp Ther 210:127
342. Yokoi, K, Murase K, Shiobara Y (1983) Life Sci 33:1665
343. Findlay JWA, Warren JT, Hill JA, Welch RM (1981) J Pharm Sci 70:624
344. Beaulieu N, Charette C, Loo JC, Jordan N, McGilveray IJ (1985) J Pharmaceut Biomed Anal 3:575
345. Cook CE, Twine ME, Myers M, Amerson E, Kepler JA, Taylor GF (1976) Res Commun Chem Pathol Pharmacol 13:497
346. Neese AL, Soyka LF (1977) Clin Pharmacol Ther 21:633
347. Raso V (1977) Cancer Treatment Rep 61:585
348. Brothman AR, Davis TP, Duffy JJ, Lindell TJ (1982) Cancer Res 42:1184
349. Van Vunakis H, Langone JJ, Riceberg LJ, Levine L (1974) Cancer Res 34:2546
350. Langone JJ, Van Vunakis H, Bachur NR (1975) Biochem Med 12:283
351. Piall E, Aherne GW, Marks V (1982) Clin Chem 28:119
352. Lewis JE, Nelson JC, Elder HA (1975) Antimicrob Agents Chemother 7:42
353. Stevens P, Young LS, Hewitt WL (1976) J Antibiotics 29:829
354. Okabayashi T, Mihara S, Repke DB, Moffat JG (1977) Cancer Res 37:619
355. Okabayashi T, Moffat JG (1982) Methods in Enzymology 84:470
356. Shimada N, Ueda T, Yokoshima T, Oh-Ishi J-I, Oh-Oka T (1984) Cancer Lett 24:173
357. Okabayashi T, Mihara S, Moffatt JG (1977) Cancer Res 37:625
358. Okabayashi T, Mihara S, Repke DB, Moffatt JG (1977) Cancer Res 37:3132
359. Broughton A, Strong JE (1976) Cancer Res 36:1418
360. Crooke ST, Luft F, Broughton A, Strong J, Casson K, Einhorn L (1977) Cancer 39:1430
361. Crooke ST, Strong JE, Broughton A (1977) Fed Proc 36:303
362. Teale JD, Clough JM, Marks V (1977) J Cancer 35:822
363. Aherne GW, James SL, Marks V (1982) Clin Chim Acta 119:341
364. Broughton A (1982) Methods in Enzymology 84:463
365. Fong K-L, Ho DHW, Carter CJK, Brown NS, Benjamin RS, Freireich EJ, Bodey GP (1980) Anal Biochem 105:281
366. Hamburger RN (1966) Science 152:203
367. Hamburger RN, Douglass JH (1969) Immunol 17:587
368. Hock C, Liemann F (1985) Arch Lebensmittelhyg 36:125
369. Gilbertson TJ, Stryd RP (1976) Clin Chem 22:828
370. Bachur NR, Riggs CE, Green MR, Langone JJ, Van Vunakis H, Levine L (1977) Clin Pharmacol Ther 21:70
371. Shimada N, Ueda T, Yokoshima T, Umemura K, Shomura T (1982) Antimicrob Agents Chemother 22:976
372. Fong, K-LL, Ho DHW, Yap BS, Stewart D, Brown NS, Benjamin RS, Freireich EJ, Bodey GP (1980) Cancer Treatment Rep 64:1253
373. Ho DH, Kanellopoulos KS, Brown NS, Issell BF, Bodey GP (1985) J Immunol Meth 85:5
374. Lewis JE, Nelson JC, Elder HA (1972) Nature New Biol 239:214
375. Mahon WA, Ezer J, Wilson JT (1973) Antimicrob Agents Chemother 3:585
376. Berk LS, Lewis JL, Nelson JC (1974) Clin Chem 20:1159
377. Minshew BH, Holmes RK, Baxter CR (1975) Antimicrob. Agents Chemother 7:107
378. Stevens P, Young LS, Hewitt WL (1975) J Lab Clin Med 86:349
379. Broughton A, Strong JE (1976) Clin Chim Acta 66:125
380. Longmore P, Atkins RC, Casley D, Johnston CI (1976) Med J Aust 1:738
381. Langone JJ (1982) Methods in Enzymology 84:409
382. Winkelhake JL, Buckmire FLA (1977) Cancer Res 37:1197
383. Jaton J-C, Ungar-Waron H (1967) Arch Biochem Biophys 122:157
384. Bohuon C, Duprey F, Boudene C (1974) Clin Chim Acta 57:263
385. Levine L, Powers E (1974) Res Commun Chem Pathol Pharmacol 9:543
386. Raso V, Schreiber R (1975) Cancer Res 35:1407
387. Hendel J, Sarek LJ, Hvidberg EF (1976) Clin Chem 22:813
388. Loeffler LJ, Blum MR, Nelsen MA (1976) Cancer Res 36:3306

389. Aherne GW, Piall EM, Marks V (1977) Br J Cancer 36:608
390. Hendel J, Sarek LJ (1977) Scand J Clin Lab Invest 37:273
391. Kamel R, Landon J, Forrest GC (1980) Clin Chem 26:97
392. Kamel RS, Landon J (1982) Methods in Enzymology 84:422
393. Tracey KJ, Mutkoski R, Lopez JA, Franzblau W, Franzblau C (1983) Cancer Chemother Pharmacol 10:96
394. Bore P, Rahmani R, Cano J-P, Just S, Barbet J (1984) Clin Chim Acta 141:135
395. Nicolau G, Szucs-Myers V, McWilliams W, Morrison J, Lanzilotti A (1985) Investigational New Drugs 3:51
396. Stevens P, Young LS, Hewitt WL (1977) Antimicrob Agents Chemother 11:768
397. Gleich GJ, Stankievic RR (1969) Immunochem 6:85
398. Karchmer AW, Dismukes WE, deRiel K, Hyslop NE (1972) Clin Res 20:531
399. Wiedermann G, Stemberger H, Werner H-P, Kraft D (1972) Immunitäts.-forsch Exp Klin Immunol 143:491
400. Hyslop NE, Calvert JF, Milligan RT, Kern K (1975) Fed Proc 34:975
401. Wal J-M, Bories G, Mamas S, Dray F (1975) FEBS Lett 57:9
402. Wal J-M, Kann G (1975) C R Acad Sci Paris 280:2405
403. Wal J-M, Bories GF (1981) Anal Biochem 114:263
404. Aherne GW, Saesow N, James S, Marks V (1982) Cancer Treatment Rep 66:1365
405. Hirnle P, Conzelmann E, Seeger H (1984) IRCS Med Sci 12:860
406. Burke JF, Mark VH, Soloway AH, Leskowitz S (1966) Cancer Res 26:1893
407. Joniau M, Stevens E, de Smet A, Verbist L (1976) Immunochem 13:715
408. Broughton A, Strong JE, Bodey GP (1976) Antimicrob Agents Chemother 9:247
409. Wedum AG (1942) J Infect Dis 70:173
410. Rubin B, Aasted B (1971) Acta Pathol Microbiol Scand 79:126
411. Gaver RC, Dixon CW, Smyth RD (1986) Cancer Chemother Pharmacol 16:207
412. Queng CT, Duber CD, McGovern J (1965) J Allergy 36:505
413. Faraj BA, Ali FM (1981) J Pharmacol Exp Ther 217:10
414. Broughton A, Strong JE, Pickering LK, Bodey GP (1976) Antimicrob Agents Chemother 10:652
415. Teale JD, Clough JM, Marks V (1977) Br J Clin Pharmacol 4:169
416. Owellen RJ, Blair M, Van Tosh A, Hains FC (1981) Cancer Treatment Rep 65:469
417. Rahmani R, Barbet J, Cano J-P (1983) Clin Chim Acta 129:57
418. Rahmani R, Martin M, Barbet J, Cano J-P (1984) Cancer Res 44:5609
419. Fasth A, Sollenberg J, Sörbo B (1975) Acta Pharm Suec 12:311
420. Wurzburger RJ, Miller RL, Boxenbaum HG, Spector S (1977) J Pharmacol Exp Ther 203:435
421. Spector S (1982) Methods in Enzymology 84:555
422. Cook CE, Ballentine N, Seltzman T, Tallent CR (1977) Pharmacologist 19:128
423. Satoh H, Hayashi M, Satoh S (1982) J Pharm Pharmacol 34:429
424. Cook CE, Kepler JA, Christensen HD (1973) Res Commun Chem Pathol Pharmacol 5:767
425. Tigelaar RE, Rapport RL, Inman JK, Kupferberg HJ (1973) Clin Chim Acta 43:231
426. Morris BA, Aherne GW, Marks V (1974) J Endocrinol 64:7
427. Cook CE, Amerson E, Poole WK, Lesser P, O'Tuama L (1975) Clin Pharmacol Ther 18:742
428. Robinson JD, Morris BA, Aherne GW, Marks V (1975) Br J Clin Pharmacol 2:345
429. Orme ML, Borga O, Cook CE, Sjöqvist F (1976) Clin Chem 22:246
430. Paxton JW, Rowell FJ, Ratcliffe JG, Lambie DG, Nanda R, Melville ID, Johnson RH (1977) Eur J Clin Pharmacol 11:71
431. Paxton JW, Rowell FJ, Ratcliffe JG (1977) Clin Chim Acta 79:81
432. Spiegel HE, Symington J, Savulich A (1978) Res Commun Chem Pathol Pharmacol 19:271
433. Ötting F (1982) Methods in Enzymology 84:577
434. Glogner P, Heni N, Nissen L (1977) Arzneim.-Forsch Drug Res 27:1703
435. Matsuda A, Kuzuya T, Sugita Y, Kawashima K (1983) Horm Metabol Res 15:425
436. Heptner W, Badian M, Baudner S, Hellstern C, Irmisch R, Rupp W, Weimer K, Wissmann H (1984) Pharm Res 5:215
437. Suzuki H, Miki M, Sekine Y, Kagemoto A, Negoro T, Maeda T, Hashimoto M (1981) J Pharm Dyn 4:217

438. Maggi E, Pianezzola E, Valzelli G (1981) Eur J Clin Pharmacol 21:251
439. Kopitar Z, Kompa HE (1975) Arzneim.-Forsch Drug Res 25:1469
440. Nieuweboer B, Gabriel D, Lübke K (1976) Arzneim.-Forsch Drug Res 26:1633
441. Yalow RS, Berson SA (1960) J Clin Invest 39:1157
442. Silbert CK, Sawin CT (1975) Clin Chem 21:1520
443. Davis JW, Yoder JM, Adams EC (1976) Clin Chim Acta 66:379
444. Wilson MA, Miles LEM (1977) Handbook of Radioimmunoassay, Abraham GE (ed), Marcel Dekker, New York and Basel, chapt 8
445. Heding LG (1969) Horm Metab Res 1:145
446. Dixon K (1974) Clin Chem 20:1275
447. Gennaro WD, Van Norman JD (1975) Clin Chem 21:873
448. Davidson P, Deal RB (1976) J Lab Clin Med 87:1050
449. Virella G, Russel D, Laimins M, Colwell J (1980) Clin Chem 26:1357
450. Palmer JP, Asplin CM, Clomons P, Lyen K, Tatpati O, Raghu PK, Paquette TL (1983) Science 222:1337
451. Collins ACG, Pickup JC (1985) Diabetic Med 2:456
452. Brodal BP (1971) Scand J Lab Clin Invest 28:287
453. Cantrell JW, Hochholzer JM, Frings CS (1972) Clin Chem 18:1423
454. Nerurkar SG, Gambhir KK (1981) Clin Chem 27:607
455. Dwenger A, Trautschold I (1982) J Clin Chem Clin Biochem 20:29
456. Dwenger A, Röllig G, Bojanovski D, Canzler H, Trautschold I (1982) Fresenius Z Anal Chem 311:358
457. Kubasik NP, Ricotta M, Hunter T, Sine HE (1982) Clin Chem 28:164
458. Livesey J, Donald RA, Roud HK (1983) Clin Chem 29:1559
459. Walters E, Henley R, Branes I (1986) Clin Chem 32:224
460. Phillips AP, Webb B, Curry AS (1972) J Forensic Sci 17:460
461. Sturner WQ, Putnam RS (1972) J Forensic Sci 17:514
462. Lindquist O (1973) Forensic Sci 2:55
463. Lindquist O, Rammer L (1975) Z Rechtsmed 75:275
464. Dickson SJ, Cairns ER, Blazey ND (1977) Forensic Sci 9:37
465. Martin FIR, Hansen N, Warne GL (1977) Med J Aust 1:58
466. Fletcher SM, Richards L, Moffat AC (1979) J Hum Vet Toxicol 21:197
467. Bauman WA, Yalow RS (1981) J Forensic Sci 26:594
468. Steel JM, Frier BM, Young RJ, Duncan LJP (1981) Lancet 354
469. Fletcher SM (1982) J Forensic Sci Soc 22:45
470. Schuck M, Liebhardt E, Tröger HD, Schuller E (1981) Beitr Ger Med 39:171
471. Jenkins D, Fletcher SM, Hoole A (1982) Med Sci Law 22:135
472. Fletcher SM (1983) J Forensic Sci Soc 23:5
473. Sturner WQ, Sullivan A (1983) J Forensic Sci 28:1016
474. Rubenstein AH, Lowy C, Welborn TA, Fraser TA (1967) Metabolism 16:234
475. Gould J, Pairent FW, Howard JM (1972) Surg Gynecol Obstet 134:815
476. Kuzuya H, Blix PM, Horwitz DL, Steiner DF, Rubenstein AH (1977) Diabetes 26:22
477. Kumar MS, Schmucher OP, Deodhar SD (1980) Amer J Clin Pathol 74:78
478. Krause U, von Erdmann B, Atzpodien W, Beyer J (1981) J Immunoassay 2:33
479. Meistas MT, Kumar MS, Schumacher OP (1981) Clin Chem 27:184
480. Rendell M (1983) Acta Diabet Lat 20:105
481. Heding LG (1977) Diabetologia 13:467
482. Cohen RM, Nakabayashi T, Blix PM, Rue PA, Shoelson SE, Root MA, Frank BH, Revers RR, Rubenstein AH (1985) Diabetes 34:84
483. Michiels M, Hendriks R, Heykants J (1977) Life Sci 21:451
484. Yano M, Nakamichi K, Yamaki T, Fukami T, Ishikawa K, Matsumoto I (1984) Chem Pharm Bull 32:1491
485. Woestenborghs R, Geuens I, Michiels M, Hendriks R, Heykants J (1986) Drug Develop Res 8:63
486. Midha KK, Rauw G, McKay G, Cooper JK, McVittie J (1984) J Pharm Sci 73:1144
487. Dixon R, Lucek R, Yu-Ying L, Colburn W, Parsonnet M (1980) Res Commun Chem Pathol Pharmacol 30:163

488. Brown K, Gardner JJ, Lockley WJS, Preston JR, Wilkinson DJ (1983) Ann Clin Biochem 20:31
489. Findlay JWA, Butz RF, Coker GG, DeAnagelis RL, Welch RM (1984) J Pharm Sci 73:1339
490. Boudene C, Duprey F, Bohuon C (1975) Biochem J 151:413
491. Ertel NH, Mittler JC, Akgun S, Wallace SL (1976) Science 193:233
492. Nelson JC, Berk LS, Lewis JE, Emori HW (1976) Clin Res 24:151A
493. Hare LE, Ditzler CA, Duggan DE (1977) J Pharm Sci 66:486
494. Cook CE, Tallent CR, Amerson E, Christensen HD, Taylor G, Kepler JA (1976) J Pharmacol Exp Ther 199:679
495. Freier C, Alberici G, Turk P, Baud F, Bohuon C (1986) Clin Chem 32:1742
496. Robins RJ, Webb AJ, Rhodes MJC, Payne J, Morgan MRA (1984) Planta Med 50:235
497. Morgan MRA, Bramham S, Webb AJ, Robins RJ, Rhodes MJC (1985) Planta Med 51:237
498. Spector S, Flynn EJ (1971) Science 174:1036
499. Flynn EJ, Spector S (1972) J Pharmacol Exp Ther 181:547
500. Cleeland R, Davis R, Heveran J, Grunberg E (1975) J Forensic Sci 20:45
501. Roerig DL, Lewand DL, Mueller MA, Wang RIH (1975) Clin Chem. 21:672
502. Jain NC, Budd RD, Sneath TC, Chinn DM, Leung WJ (1976) Clin Tox 9:221
503. Budd RD, Yang FC, Utley KO (1981) Clin Tox 18:317
504. Law B, Moffat AC (1981) J Forensic Sci Soc 21:55
505. Mason PA, Law B (1982) J Label Comp Radiopharm 19:357
506. Takatori T, Terazawa K (1982) Japan J Legal Med 36:414
507. Flynn EJ, Spector S (1974) J Pharmacol Exp Ther 189:550
508. Chung A, Kim SY, Cheng LT, Castro A (1973) Experientia 29:820
509. Satoh H, Kuroiwa Y, Hamada A (1973) J Biochem 73:1115
510. Satoh H, Kuroiwa Y (1974) J Biochem 76:1293
511. Satoh H, Kuroiwa Y, Hamada A, Uematsu T (1974) J Biochem 75:1301
512. Viswanathan CT, Booker HE, Welling PG (1977) Clin Chem 23:873
513. Yamaoka A, Takatori T (1979) J Immunol Meth 28:51
514. Aderjan R, Schmidt G (1979) Z Rechtsmed 83:191
515. Sherman-Gold R, Dudai Y, Fogelfeld L, Fuchs S (1983) J Immunoassay 4:135
516. Jacqmin P, Lesne M (1984) J Pharm Belg 39:5
517. Goddard CP, Law B, Mason PA, Stead AH (1986) J Label Comp Radiopharm 23:383
518. Goddard CP, Stead AH, Mason PA, Law B, Moffat AC, McBrien M, Cosby S (1986) Analyst 111:525
519. Bechtel WD, Weber KH (1985) J Pharm Sci 74:1265
520. Dixon WR, Earley J, Postma E (1975) J Pharm Sci 64:937
521. Dixon R, Lucek R, Earley J, Perry C (1979) J Pharm Sci 68:261
522. Lucek R, Dixon R (1980) Res. Commun Chem Pathol Pharmacol 27:397
523. Dixon R (1982) Methods in Enzymology 84:490
524. Dixon WR, Young RL, Ning R, Liebman A (1975) Pharmacologist 17:251
525. Dixon WR, Young RL, Ning R, Liebman A (1977) J Pharm Sci 66:235
526. Rosenthaler J, Nimmerfall F, Sigrist R, Munzer H (1977) Eur J Biochem 80:603
527. Peskar B, Spector S (1973) J Pharmacol Exp Ther 186:167
528. Gelbke HP, Schlicht HJ, Schmidt G (1977) Arch Toxicol 38:295
529. Bourne RC, Robinson JD, Teale JD (1978) Br J Pharmacol 63:371P
530. Dixon R, Crews T (1978) J Analyt Toxicol 2:210
531. Gelbke HP, Schlicht HJ, Schmidt G (1978) Z Rechtsmed 80:319
532. Dixon R, Glover W, Earley J (1981) J Pharm Sci 70:230
533. Glover W, Dixon R (1979) Fed Proc 38:443
534. Glover W, Earley J, Delaney M, Dixon R (1980) J Pharm Sci 69:601
535. Ko H, Royer ME, Hester JB, Johnston KT (1977) Analyt Lett 10:1019
536. Goldstein SA, Van Vunakis H (1981) J Pharmacol Exp Ther 217:36
537. Peinhardt G, Hergert I, Giessler AJ (1984) Pharmazie 39:37
538. Hergert I, Hecht HF, Peinhardt G (1986) Pharmazie 41:289
539. Shostak M (1974) The Phenothiazines and Structurally Related Drugs, Forrest IS, Carr CJ, Usdin E (eds), Raven Press, New York, p 365
540. Shostak, M (1974) Adv Biochem Psychopharmacol 9:365

541. Spector S (1974) The Phenothiazines and Structurally Related Drugs, Forrest IS, Carr CJ, Usdin E (eds), Raven Press, New York, p 363
542. Kawashima K, Dixon R, Spector S (1975) Eur J Pharmacol 32:195
543. Midha KK, Loo JCK, Charette C, Rowe ML, Hubbard JW, McGilveray IJ (1978) J Anal Toxicol 2:185
544. Midha KK, Loo JCK, Hubbard JW, Rowe ML, McGilveray IJ (1979) Clin Chem 25:166
545. Grassi J, Nevers MC, Pradelles P (1981) Pathol Biol 29:375
546. Yeung PKF, Hubbard JW, Cooper JK, Midha KK (1983) J Pharmacol Exp Ther 226:833
547. Robinson JD, Risby D (1977) Clin Chem 23:2085
548. Jørgensen A (1978) Life Sci 23:1533
549. Rowell FJ, Hui SM, Paxton JW (1979) J Immunol Meth 31:159
550. Wiles DH, Gelder MG (1979) Br J Clin Pharmacol 8:565
551. Midha KK, Cooper JK, Hubbard JW (1980) Commun Psychopharmacol 4:107
552. Hawes EM, Shetty HU, Cooper JK, Rauw G, McKay G, Midha KK (1984) J Pharm Sci 73:247
553. Midha KK, MacKonka C, Cooper JK, Hubbard JW, Yeung PFK (1981) Br J Clin Pharmacol 1:85
554. Midha KK, Hawes EM, Rauw G, McVittie J, McKay G, Cooper JK, Shetty HU (1983) Ther Drug Monit 5:117
555. Midha KK, Hubbard JW, Cooper JK, Hawes EM, Fournier S, Yeung P (1981) Br J Clin Pharmacol 12:189
556. Hawes EM, Aravagiri M, Dulos RA, Rauw GA, Stonkus MD (1983) Prog Neuro-Psychopharmacol Biol Psychiat 7:709
557. Midha KK, Hawes EM, Rauw G, McKay G, Cooper JK, Shetty HU (1983) J Pharmacol Meth 9:283
558. Aravagiri M, Hawes EM, Midha KK (1985) J Pharm Sci 74:1196
559. Aravagiri M, Hawes EM, Midha KK (1986) J Pharmacol Exp Ther 237:615
560. Aravagiri M, Hawes EM, Midha KK (1984) J Pharm Sci 73:1383
561. McKay G, Rauw GAJ, Stonkus MD, Dulos RA, Gedir RG, Hawes EM, Midha KK (1984) J Pharmacol Meth 12:203
562. Kaul B, Quame B, Davidow B (1977) J Anal Toxicol 1:236
563. Buchman O, Pri-Bar I (1978) J Label Comp Radiopharm 14:263
564. Hubbard JW, Midha KK, Cooper JK, Charette C (1978) J Pharm Sci 67:1571
565. Maguire KP, Burrows GD, Norman TR, Scoggins BA (1978) Clin Chem 24:549
566. Aherne GW, Marks V, Mould G, Stout G (1977) Lancet 1214
567. Lucek R, Dixon R (1977) Res Commun Chem Pathol Pharmacol 18:125
568. Braithwaite RA, Montgomery S, Robinson JD (1978) Br J Pharmacol 63:370P
569. Robinson JD, Risby D, Riley G, Aherne GW (1978) J Pharmacol Exp Ther 205:499
570. Brunswick DJ, Needelman B, Mendels J (1979) Br. J Clin Pharmacol 7:343
571. Virtanen R (1980) Scand J Clin Lab Invest 40:191
572. Goyot C, Golse B, Grenier J (1981) Pathol Biol 29:330
573. Goyot C, Grenier J, Hug R (1983) Sem Hôp Paris 59:350
574. Read GF, Riad-Fahmy D, Walker RF (1977) Postgrad Med J 53:110
575. Read GF, Riad-Fahmy D (1978) Clin Chem 24:36
576. Spector S, Spector NL, Almeida MP (1975) Psychopharmacol Commun 1:421
577. Brunswick DJ, Needelman B, Mendels J (1978) Life Sci 22:137
578. Midha KK, Charette C (1980) Commun Psychopharmacol 4:11
579. Virtanen R, Salonen JS, Scheinin M, Iisalo E, Mattila V (1980) Acta Pharmacol Toxicol 47:274
580. Robinson JD, Braithwaite RA, Dawling S (1978) Clin Chem 24:2023
581. Goyot G, Adeline J, Grenier J (1983) Pathol Biol 31:654
582. Sayegh JF (1986) Neurochem Res 11:193
583. Butz RF, Schroeder DH, Welch RM, Mehta NB, Phillips AP, Findlay JWA (1981) J Pharmacol Exp Ther 217:602
584. Butz RF, Smith PG, Schroeder DH, Findlay JWA (1983) Clin Chem 29:462
585. Mehta NB, Musso DL (1986) J Pharm Sci 75:410
586. Shostak M, Perel JM (1976) Fed Proc 35:531

587. Clark BR, Tower BB, Rubin RT (1977) Life Sci 20:319
588. Rubin RT, Tower BB, Hays SE, Poland RE (1982) Methods in Enzymology 84:532
589. Browning JL, Harrington CA, Davis CM (1985) J Immunoassay 6:45
590. Berman AR, McGrath JP, Permisohn RC, Cella JA (1975) Clin Chem 21:1878
591. Bost RO, Sutheimer CA, Sunshine I (1976) Clin Chem 22:689
592. Kogan MJ, Jukofsky D, Verebey K, Mulé SJ (1978) Clin Chem 24:1425
593. Mulé SJ, Kogan M, Jukofsky D (1978) Clin Chem 24:1473
594. Heptner W, Badian MJ, Baudner S, Christ OE, Fraser HM, Rupp W, Weimer KE, Wissmann H (1977) Br J Pharmacol 4:123S
595. Michiels M, Hendriks R, Heykants J (1982) Methods in Enzymology 84:542
596. Mizuchi A, Saruta S, Kitigawa N, Miyachi Y (1981) Arch Int Pharmacodyn 254:317
597. Cardoso MT, Pradelles P (1982) J Label Comp Radiopharm 14:1103
598. Cooper DS, Bode HH, Nath B, Saxe V, Maloof F, Ridgway EC (1984) J Clin Endocrinol Metab 58:473
599. Halpern R, Cooper DS, Kieffer JD, Saxe V, Mover H, Maloof F, Ridgway EC (1983) Endocrinol 113:915
600. Schwenk R, Kelly K, Tse KS, Sehon AH (1975) Clin Chem 21:1059
601. Tse KS, Vijay H, Attallah N, Sehon AH (1976) J Immunol 116:965
602. Peskar BM, Peskar BA, Turner JC (1976) J Pharm Pharmacol 28:720
603. Bozler G (1977) Proc Int Symp 2:299
604. Fellows I, Jenner WN, Martin LE, Willoughby BA (1981) Br J Pharmacol 73:274P
605. Jenner WN, Martin LE, Willoughby BA, Fellows I (1981) Life Sci 28:1323
606. Quinn RP, Scharver J, Hill JA (1985) Pharmac Ther 30:43
607. Quinn RP, de Miranda P, Gerald L, Good SS (1979) Anal Biochem 98:319
608. Skubitz KM, Quinn RP, Lietman PS (1982) Antimicrob Agents Chemother 21:352
609. Quinn RP, Good SS, Gerald L, Sabatka JJ (1983) Anal Biochem 134:16
610. Quinn RP, Tadepalli SM, Gerald L (1985) Fed Proc 44:1122
611. Piall EM, Aherne GW, Marks VM (1979) Br J Cancer 40:548
612. Austin RK, Trefts PE, Hintz M, Connor JD, Kagnoff MF (1983) Antimicrob Agents Chemother 24:696
613. Nerenberg C, Tsina I, Matin S (1982) J Assoc Off Anal Chem 65:635
614. Gourmel B, Fiet J, Collins RF, Villette JM, Dreux C (1980) Clin Chim Acta 108:229
615. Gourmel B, Fiet J, Collins RF, Villette JM, Passa P, Dreux C (1981) Clin Chim Acta 115:229
616. Rattenberger E, Matzke P, Neudegger J (1985) Arch Lebensmittelhyg 36:85
617. Chu S-T, Vega SM, Ali A, Sennello LT (1981) J Pharm Sci 70:990
618. Kawashima K, Levy A, Spector S (1976) J Pharmacol Exp Ther 196:517
619. Marks V, Mould GP, Stout S, Williams S (1978) Br J Clin Pharmarcol 5:371P
620. Mould GP, Clough J, Morris BA, Stout G, Marks V (1981) Biopharmaceut Drug Disposit 2:49
621. Eller TD, Knapp DR, Walle T (1983) Anal Chem 55:1572
622. Lesne M, Dolphen R (1977) J Immunol Meth 17:189
623. Warren JT, Coker GG, Welch RM, Fowle ASE, Findlay JWA (1982) J Pharm Sci 71:665
624. Gold EF, Ben-Efraim S, Faivisewitz A, Steiner Z, Klajman A (1977) Clin Immunol Immunopathol 7:176
625. Russell AS, Ziff M (1968) Clin Exp Immunol 3:901
626. Cook CE, Tallent CR, Amerson E, Taylor G, Kepler JA (1975) Pharmacologist 17:219
627. Duncan FM, Martin VI, Williams BC, Al-Dujaili EAS, Edwards CRW (1983) Clin Chim Acta 131:295
628. Tu J, Liu E, Nickoloff EL (1984) Therapeutic Drug Monit 6:59
629. Jarrott B, Spector S (1977) Fed Proc 36:949
630. Jarrott B, Spector S (1978) J Pharmacol Exp Ther 207:195
631. Arndts D, Stähle H, Struck CJ (1979) Arzneim.-Forsch/Drug Res 29:532
632. Arndts D, Stähle H, Förster H-J (1981) J Pharmacol Meth 6:295
633. Farina PR, Homon CA, Chow CT, Keirns JJ, Zavorskas PA, Esber HJ (1985) Therapeutic Drug Monit 7:344
634. Vakkuri O, Leppäluoto J, Vuolteenaho O, Järvensivu P, Karjalainen A, Kukela K (1984) Scand J Lab Clin Invest 44:603

635. Worland PJ, Jarrott B (1986) J Pharm Sci 75:512
636. Ohata K, Yokoyama N, Sakamoto H, Kohno S, Kobayashi M, Murai K, Kashima K, Tatsumi H (1984) Arzneim.-Forsch/Drug Res 34:1704
637. Ellman L, Inman J, Green I (1971) Clin Exp Immunol 9:927
638. Yamauchi Y, Litwin A, Adams L, Zimmer H, Hess EV (1975) J Clin Invest 56:958
639. Levy A, Kawashima K, Spector S (1976) Life Sci 19:1421
640. Pettinger WA, Keeton K, Phil M, Tanaka K (1975) Clin Pharmacol Ther 17:146
641. Pettinger WA, Mitchell HC (1976) Fed Proc 35:2521
642. Arndts D, Stähle H (1981) J Pharmacol Meth 6:109
643. Flasch H (1975) Klin Wochenschr 53:873
644. Plum J, Daldrup T (1985) Z Rechtsmed 94:257
645. Selden R, Klein MD, Smith TW (1973) Circulation 47:744
646. Smith TW (1972) J Clin Invest 51:1583
647. Fletcher SM, Lawson G, Law B, Moffat AC (1980) J Forensic Sci Soc 20:203
648. Brooker G, Fleming JW, Ross EM (1982) Biochim Biophys Acta 685:39
649. Boguslaski RC, Schwartz CL (1975) Anal Chem 47:1583
650. Stoll RG, Christensen MS, Sakmar E, Blair D, Wagner JG (1973) J Pharm Sci 62:1615
651. Oliver GC, Parker BM, Brasfield DL, Parker CW (1968) J Clin Invest 47:1035
652. Smith TW (1970) J Pharmacol Exp Ther 175:352
653. Hansteen V, Jacobsen D, Knudsen K, Reikvam A, Skuterud B (1981) Clin Toxicol 18:679
654. Plum J, Daldrup T (1986) J Chromatog 377:221
655. Smith TW, Butler VP, Haber E (1970) Biochem 9:331
656. Horgan ED, Riley WJ (1973) Clin Chem 19:187
657. Blazey ND (1977) Clin Chim Acta 80:403
658. Smith TW, Butler VP, Haber E (1969) New England J Med 281:1212
659. Oliver GC, Parker BM, Parker CW (1971) Amer J Med 51:186
660. Line WF, Siegel SJ, Kwong A, Frank C, Ernst R (1973) Clin Chem 19:1361
661. Phillips AP (1973) Clin Chim Acta 44:333
662. Smith TW, Haber E (1973) Pharmacol Rev 25:219
663. Boguslaski RC, Denning CE (1975) Biochem Med 14:83
664. Loo JCK, McGilveray IJ, Jordan N (1981) J Liq Chromatog 4:879
665. Butler VP, Tse-Eng D (1982) Methods in Enzymology 84:558
666. Gault MH, Longerich L, Dawe M, Vasdev SC (1985) Clin Chem 31:1272
667. Moore RA, Turner M, Cousins J, Gales M (1985) Ann Clin Biochem 22:190
668. Pudek MR, Seccombe DW, Jacobson BE, Whitfield MF (1983) Clin Chem 29:1972
669. D'Arcy PF (1984) Pharmacy Int. 188
670. Balzan S, Clerico A, del Chicca MG, Montali U, Ghione S (1984) Clin Chem 30:450
671. Koren G, Farine D, Maresky D, Taylor J, Heyes J, Soldin S, MacLeod S (1984) Clin Pharmacol Ther 36:759
672. Vinge E, Nilsson L-G, Molin L, Ekman R (1984) New England J Med 310:725
673. Greenway DC, Nanji AA (1985) Clin Chem 31:1078
674. McCarthy RC (1985) Clin Chem 31:1240
675. Oldfield PR, Singer BE, Toseland PA (1985) Clin Chem 31:1246
676. Pudek MR, Seccombe DW, Jacobson BE, Humphries K (1985) Clin Chem 31:1806
677. Valdes R (1985) Clin Chem 31:1525
678. Pleasants RA, Gadsden RH, McCormack JP, Piveral K, Sawyer WT (1986) Clin Pharm 5:810
679. Pudek MR, Seccombe DW, Humphries K (1986) Clin Chem 32:2005
680. Soldin SJ (1986) Clin Chem 32:5
681. Cerceo E, Elloso CA (1972) Clin Chem 18:539
682. Belpaire FM, Bogaert MG, de Broe NE (1975) Clin Chim Acta 62:255
683. Kubasik NP, Brody BB, Barold SS (1975) Amer J Cardiol 36:975
684. Kuno-Sakai H, Sakai H, Ritzmann SE (1975) Clin Chem 21:156
685. Padeletti L (1975) J Nucl Biol Med 19:22
686. Ravel R (1975) Clin Chem 21:1801
687. Kramer WG, Bathala MS, Reuning MH (1976) Res Commun Chem Pathol Pharmacol 14:83

688. Lindenbaum J, Maulitz RM, Butler VP (1976) Gastroenterol 71:399
689. Boone, J, Griffin C, Shaw W (1977) Clin Chem 23:2180
690. Müller H, Bräuer H, Resch B (1978) Clin Chem 24:706
691. DiPiro JT, Cote JR, DiPiro CR, Bustrack CA (1980) Amer J Hosp Pharm 37:1518
692. Scherrmann JM, Bourdon R (1980) Clin Chem 26:670
693. Berdeaux A, Valette H, Auzepy P, Blondeau M, Giudicelli JF (1981) Nouv Presse Méd 10:3650
694. Bergdähl B, Molin L (1981) Clin Biochem 14:67
695. Gibb I, Adams PC, Parnham AJ, Jennings K (1983) Br J Clin Pharmacol 16:445
696. Stockley J (1983) Pharmaceut J 230:34
697. Smith TW, Willerson JT (1971) Circulation 44:29
698. Coltart J, Howard M, Chamberlain D (1972) Br Med J 318
699. Hobson JD, Zettner A (1973) J Amer Med Assoc 223:147
700. Iisalo E, Nuutila M (1973) Lancet 257
701. Brock A (1974) Acta Pharmacol Toxicol 34:198
702. Karjalainen J, Ojala K, Reissell P (1974) Acta Pharmacol Toxicol 34:385
703. Moffat AC (1974) Acta Pharmacol Toxicol 35:386
704. Phillips AP (1974) J Forensic Sci 19:900
705. Phillips AP (1974) J Forensic Sci Soc 14:137
706. Di Maio VJM, Garriott JC, Putnam R (1975) J Forensic Sci 20:340
707. Holt DW, Benstead JG (1975) J Clin Path 28:483
708. Kim PW, Krasula RW, Soyka LF, Hastreiter AR (1975) Circulation 52:1128
709. Selesky M, Spiehler V, Cravey RH, Elliot HW (1976) J Forensic Sci 22:409
710. Dickson SJ, Blazey ND (1977) Forensic Sci 9:145
711. Vorpahl TE, Coe JI (1978) J Forensic Sci 23:329
712. Aderjan R, Buhr H, Schmidt G (1979) Arch Toxicol 42:107
713. Fletcher SM, Lawson GJ, Moffat AC (1979) J Forensic Sci Soc 19:183
714. Sedgwick P, Spiehler VR, Cravey RH (1981) Clin Tox 18:887
715. Smith PF (1981) J Forensic Sci 26:193
716. Spiehler VR, Sedgwick P, Richards RG (1981) J Forensic Sci 26:645
717. Osterich J, Herold S, Pond S (1982) J Amer Med Assoc 247:1596
718. Hastreiter AR, Kim PW, van der Horst R (1983) J Forensic Sci 28:482
719. Ng RH, Stempsey W, Statland BE (1983) Clin Chem 29:393
720. Hastreiter AR, van der Horst RL (1984) J Forensic Sci 29:139
721. Lang D, Belz GG, Hofstetter R, Töllner U, von Bernuth G (1976) Klin Wschr 54:389
722. Kawai S, Ogawa K, Satake T (1982) Clin Pharmacol Ther 31:541
723. Joubert PH, Müller FO, Aucamp BN (1976) Brit J Clin Pharmacol 3:673
724. Greenwood H, Snedden W, Hayward RP, Landon J (1975) Clin Chim Acta 62:213
725. Christenson RH, Hammond JE, Hull JH, Bustrack JA (1982) Clin Chim Acta 120:13
726. O'Brien MS, Huston JR, Reid RJ, Gibson TP (1985) Clin Chem 31:122
727. Marcus FI, Ryan JN, Stafford MG (1975) J Lab Clin Med 85:610
728. Flasch H, Heinz N, Petersen R (1977) Arzneim.-Forsch 27:649
729. Loo JCK, McGilveray IJ, Jordan N (1977) Res Commun Chem Pathol Pharmacol 17:497
730. Kramer WG, Bathala MS, Reuning RH (1976) Res Commun Chem Pathol Pharmacol 14:83
731. Butler VP, Tse-Eng D, Lindenbaum J, Kalman SM, Preibisz J, Rund DG, Wissel PS (1982) J Pharmacol Exp Ther 221:123
732. Lesne M (1972) Arch Int Pharmacol Thér 199:206
733. Lesne M, Dolphen R (1976) J Pharmacol (Paris) 7:619
734. Karjalainen J, Ojala K (1975) Klin Wschr 53:685
735. Bodem G, Ochs H, Hahn E, Dengler HJ (1977) Schweiz Med Wschr 107:658
736. Hartel G, Manninen V, Melin J, Apajalahti A (1973) Ann Clin Res 5:87
737. Boerner D, Olcay A, Schaumann W, Weiss W (1976) Eur J Clin Pharmacol 9:307
738. Johnson BF, Bye CE, Jones GE, Sabey GA (1976) Eur J Clin Pharmacol 10:231
739. Rietbrock N, Guggenmos J, Kuhlmann J, Hess U (1976) Eur J Clin Pharmacol 9:373
740. Garrett ER, Hinderling PH (1977) J Pharm Sci 66:806
741. Marinow J, Olcay A, Schaumann W, Weiss W (1977) Eur J Clin Pharmacol 11:213

742. Arnold W, Püschel K (1979) Z Rechtsmed 83:265
743. Selden R, Smith TW (1972) Circulation 45:1176
744. Smith TW (1972) J Clin Invest 51:1683
745. Belz GG, Brech WJ, Kleeberg UR, Rudofsky G, Belz G (1973) Naunyn-Schmiedeberg's Arch Pharmacol 279:105
746. Mulé SJ, Whitlock E, Jukofsky D (1975) Clin Chem 21:81
747. Bost RO, Sutheimer CA, Sunshine I (1976) Clin Chem 22:789
748. Faraj BA, Israili ZH, Knight NE, Smissman EE, Pazdernik TJ (1976) J Med Chem 19:20
749. Budd RD (1981) Clin Toxicol 18:299
750. Smith FP (1981) Forensic Sci 17:225
751. Budd RD (1982) J Chromatog 245:129
752. Reynolds GP, Elsworth JD, Blau K, Sandler M, Lees AJ, Stern GM (1978) Br J Clin Pharmacol 6:542
753. Reynolds EF, Prasad AB (eds) (1982) Martindale. The Extra Pharmacopeia. 28th ed, The Pharmaceutical Press, London, p 1632
754. Cook CE, Tallent CR, Amerson EW, Myers MW, Kepler JA, Taylor GF, Christensen HD (1976) J Pharmacol Exp Ther 199:679
755. Grant JD, Gross SJ, Lomax P, Wong R (1972) Nature New Biology 236:216
756. Teale JD, Forman EJ, King LJ, Marks V (1974) Nature 249:154
757. Gross SJ, Soares JR, Wong S-LR, Schuster RE (1974) Nature 252:581
758. Teale JD, Forman EJ, King LJ, Marks V (1974) Lancet 2:553
759. Teale JD, Forman EJ, King LJ, Marks V (1974) Proc Soc Anal Chem 11:219
760. Tsui PT, Kelly KA, Ponpipom MM, Strahilevitz M, Sehon AH (1974) Can J Biochem 52:252
761. Marks V, Teale D, Fry D (1975) Br Med J 3:348
762. Soares JR, Gross SJ (1976) Life Sci 19:1711
763. Law B, Williams PL, Moffat AC (1979) Vet Hum Toxicol 21:144
764. Williams PL, Moffat AC, King LJ (1979) J Chromatog 186:595
765. Pitt CG, Seltzman HH, Setzer SR, Williams DL (1980) J Labell Comp Radiopharm 17:681
766. Bergman RA, Lukaszewski T, Wang SYS (1981) J Analyt Toxicol 5:85
767. Owens SM, McBay AJ, Reisner HM, Perez-Reyes M (1981) Clin Chem 27:619
768. Yeager EP, Goebelsmann U, Soares JR, Grant JD, Gross SJ (1981) J Analyt Toxicol 5:81
769. Law B, Mason PA, Moffat AC, King LJ (1982) J Labell Comp Radiopharm 19:915
770. Cais M, Dani S, Shimoni M (1983) Arch Toxicol Suppl 6:105
771. Hanson VW, Buonarati MH, Baselt RC, Wade NA, Yep C, Biasotti AA, Reeve VC, Wong AS, Orbanowsky MW (1983) J Analyt Toxicol 7:96
772. Cook CE, Schindler VH, Tallent CR, Seltzman HH, Warick C, Pitt CG (1984) The Cannabinoids: Chemical, Pharmacologic, and Therapeutic Aspects, Cook CE (ed), Academic Press, New York and London, p 135
773. Childs PS, McCurdy HH (1984) J Analyt Toxicol 8:220
774. Irving J, Leeb B, Foltz RL, Cook CE, Bursey JT, Willette RE (1984) J Analyt Toxicol 8:192
775. Jones AB, ElSohly HN, Arafat ES, ElSohly MA (1984) J Analyt Toxicol 8:249
776. Jones AB, ElSohly HN, ElSohly MA (1984) J Analyt Toxicol 8:252
777. Law B, Mason PA, Moffat AC, King LJ (1984) J Analyt Toxicol 8:14
778. Law B, Mason PA, Moffat AC, King LJ (1984) J Analyt Toxicol 8:19
779. Peat MA, Deyman ME, Johnson JR (1984) J Forensic Sci 29:110
780. Frederick DL, Green J, Fowler MW (1985) J Analyt Toxicol 9:116
781. Abercrombie ML, Jewell JS (1986) J Analyt Toxicol 10:178
782. ElSohly MA, Jones AB, ElSohly HN, Stanford DF (1985) J Analyt Toxicol 9:190
783. Teale D, Marks V (1976) Lancet 1:884
784. Garrett CPO, Braithwaite RA, Teale JD (1977) Br Med J 2:166
785. Teale JD, Clough JM, King LJ, Marks V (1977) J Forensic Sci Soc 17:177
786. Law B (1981) J Forensic Sci Soc 21:31
787. Reeve VC, Robertson WB, Grant J, Soares JR, Zimmermann E, Gillespie HK, Hollister LE (1983) J Forensic Sci 28:963
788. Zimmermann EG, Yeager EP, Soares JR, Hollister LE, Reeve VC (1983) J Forensic Sci 28:957
789. Law B, Mason PA, Moffat AC, Gleadle RI, King LJ (1984) J Pharm Pharmacol 36:289

790. Mason AP, Perez-Reyes M, McBay AJ (1983) J Analyt Toxicol 7:172
791. Perez-Reyes M, Guiseppi SD, Mason AP, Davis KH (1983) Clin Pharmacol Ther 34:36
792. Law B, Mason PA, Moffat AC, King LJ, Marks V (1984) J Pharm Pharmacol 36:578
793. Mørland J, Bugge A, Skuterud B, Steen A, Wethe GH, Kjeldsen T (1985) J Forensic Sci 30:997
794. Gross SJ, Worthy TE, Nerder L, Zimmermann EG, Soares JR, Lomax P (1985) J Analyt Toxicol 9:1
795. Kaul B, Millian SJ, Davidow B (1976) J Pharmacol Exp Ther 199:171
796. Johns ME, Berman AR, Price JC, Pillsbury RC, Henderson RL (1977) Ann Otol 86:342
797. Mulé SJ, Jukofsky D, Kogan M, De Pace A, Verebey K (1977) Clin Chem 23:796
798. Budd RD (1981) Clin Toxicol 18:773
799. Fletcher SM, Hancock VS (1981) J Chromatog 206:193
800. Baumgartner WA, Black CT, Jones PF, Blahd WH (1982) J Nucl Med 23:790
801. Liu Y, Budd RD, Griesemer EC (1982) J Chromatog 248:318
802. Baselt RC (1983) J Chromatog 268:502
803. Spiehler VR, Reed D (1985) J Forensic Sci 30:1003
804. Van Vunakis H, Bradvica H, Benda P, Levine L (1969) Biochem Pharmacol 18:393
805. Kido Y, Nagamatsu K, Ishizeki C (1974) Yakugaku Xasshi 94:1290
806. Riceberg LJ, Van Vunakis H, Levine L (1974) Analyt Biochem 60:551
807. Van Vunakis H, Farrow JT, Gjika HB, Levine L (1971) Proc Nat Acad Sci 68:1483
808. Lopatin DE, Voss EW (1974) Immunochem 11:285
809. Castro A, Grettie DP, Bartos F, Bartos D (1973) Res Commun Chem Pathol Pharmacol 6:879
810. Loeffler LJ, Pierce J (1973) J Pharm Sci 11:1817
811. Taunton-Rigby A, Sher SE, Kelley PR (1973) Science 181:165
812. Wingaleth DC, Makowski A, Teitelbaum DT (1974) Clin Chem 20:870
813. Ratcliffe WA, Fletcher SM, Moffat AC, Ratcliffe JG, Harland WA, Levitt TE (1977) Clin Chem 23:169
814. Twitchett PJ, Fletcher SM, Sullivan AT, Moffat AC (1978) J Chromatog 150:73
815. Peel HW, Boynton AL (1980) Can Soc Forensic Sci J 13:23
816. Fysh RR, Oon MCH, Robinson KN, Smith RN, White PC, Whitehouse MJ (1985) Forensic Sci Int 28:109
817. Smith RN, Robinson K (1985) Forensic Sci Int 28:229
818. Stead AH, Watton J, Goddard CP, Patel AC, Moffat AC (1986) Forensic Sci Int 32:49
819. Faraj BA, Israili ZH, Knight NE, Smissman EE, Pazdernik TJ (1976) J Med Chem 19:20
820. Inayama S, Tokunaga Y, Hosoya E, Nakadate T, Niwaguchi T (1977) Chem Pharm Bull 25:840
821. Inayama S, Tokunaga Y, Hosoya E, Nakadate T, Niwaguchi T, Aoki K, Saito S (1980) Chem Pharm Bull 28:2779
822. Inoue T, Kanda Y, Kishi T, Sakai T, Suzuki S, Niwaguchi T, Hori H, Inayama S (1984) Chem Pharm Bull 32:344
823. Schnoll SH, Vogel WH, Odstrchel G (1973) Fed Proc 32:719
824. Utzinger R (1975) Psychopharmacol Berl 41:301
825. Heveran JE, Anthony M, Ward C (1980) J Forensic Sci 25:79
826. Heveran JE, Ward C (1980) J Forensic Sci 25:719
827. Kaul B, Davidow B (1980) Clin Toxicol 16:7
828. Kaul B, Davidow B (1980) J Clin Pharmacol 20:500
829. Ward DP, Trevor AJ (1980) Life Sci 27:457
830. Budd RD (1981) Clin Toxicol 18:1033
831. Budd RD, Leung WJ (1981) Clin Toxicol 18:85
832. Owens SM, Woodworth J, Mayersohn M (1982) Clin Chem 28:1509
833. Weingarten HL, Trevias EC (1982) J Analyt Toxicol 6:88
834. Budd RD (1984) J Chromatog 295:492
835. McCarron MM, Walberg CB, Soares JR, Gross SJ, Baselt RC (1984) J Analyt Toxicol 8:197
836. Hooker SB, Boyd WC (1940) J Immunol 38:479
837. Meikle AW (1982) Methods in Enzymology 84:585

838. Dixon WR, Young RL, Holazo A, Jack ML, Weinfeld RE, Alexander K, Liebman A, Kaplan SA (1976) J Pharm Sci 65:701
839. Heptner W, Baudner S, Dagrosa EE, Hellstern C, Irmisch R, Strecker H, Wissmann H (1984) J Immunoassay 5:13
840. Dray F (1983) Br J Dermatol 109:Suppl. 25, 36
841. Salmon JA (1983) Br Med Bull 39:227
842. Moonen P, Klok G, Keirse MJNC (1985) Prostaglandins 29:443
843. Young RN, Kakushima M, Rokach J (1982) Prostaglandins 23:603
844. Aharony D, Dobson P, Berstein PR, Kusner EJ, Krell RD, Smith JB (1983) Biochem Biophys Res Commun 117:574
845. Hayes EC, Lombardo DL, Girard Y, Maycock AL, Rokach J, Rosenthal AS, Young RN, Egan RW, Zweerink HJ (1983) J Immunol 131:429
846. Young RN, Zamboni R, Rokach J (1983) Prostaglandins 26:605
847. Rokach J, Hayes EC, Girard Y, Lombardo DL, Maycock AL, Rosenthal AS, Young RN, Zamboni R, Zweerink HJ (1984) Prostagland Leuk Med 13:21
848. Wynalsa MA, Brashler JR, Bach MK, Morton DR, Fitzpatrick FA (1984) Anal Chem 56:1862
849. Zijlstra F, Vincent JE (1984) J Chromatog. 311:39
850. Jaffe BM, Smith JW, Newton WT, Parker CW (1971) Science 171:494
851. Levine L (1973) Pharmacol Rev 25:293
852. Maclouf J, Pradel M, Pradelles P, Dray F (1976) Biochim Biophys Acta 421:139
853. Lahera V, Durán F, Cachofeiro V, del Cañizo FJ, Rodriguez FJ, Tresguerres JAF (1986) Rev Espan Fisiol 42:233
854. Kelly RW, Deam S, Cameron MJ, Seamark RF (1986) Prostagland Leuk Med 24:1
855. Uderman HD, Workman RJ (1986) Prostagland Leuk Med 24:69
856. Carty TJ, Falkner FC, Schaaf TK (1982) Prostagland Leuk Med 9:193
857. Nieuweboer B, Schmidt-Gollwitzer K, Schubert R, Seemann G (1982) Prostagland Leuk Med 9:183
858. Kleimola TT (1978) Br J Clin Pharmacol 6:255
859. Rosenthaler J, Munzer H (1976) Experientia 32:234
860. Rosenthaler J, Munzer H, Voges R, Andres H, Gull P, Bolliger G (1984) Int J Nucl Med Biol 11:85
861. Inada S, Kajii K, Takenaga M, Fujita M, Kawamura K, Yanaihara N (1982) Radioisotopes 31:190
862. Povšič L, Herzog B, Kopitar H (1983) Period Biol 85:Suppl. 3, 251
863. Loh W, Woodcock BG (1983) Arzneim.-Forsch/Drug Res 31:568
864. Collignon F, Pradelles P (1984) Eur J Nucl Med 9:23
865. Hümpel M, Nieuweboer B, Hasan SH, Wendt H (1981) Eur J Clin Pharmacol 20:47
866. Arens H, Zenk MH (1980) Planta Med 39:336
867. Langone JJ, Van Vunakis H (1975) Res Commun Chem Pathol Pharmacol 10:163
868. Centeno ER, Johnson WJ, Sehon AH (1970) Int Arch Allergy 37:1
869. Knopp D, Nuhn P, Dobberkau H-J (1985) Arch Toxicol 58:27
870. Haas GJ, Guardia EJ (1968) Proc Exp Biol Med 129:546
871. Levitt T (1977) Lancet 2:358
872. Levitt T (1979) Proc Analyt Div Chem Soc 16:72
873. Proudfoot AT, Stewart MS, Levitt T, Widdop B (1979) Lancet 330
874. Stewart MJ, Levitt T, Jarvie DR (1979) Clin Chim Acta 94:253
875. Ercegovich CD, Vallejo RP, Gettig RR, Woods L, Bogus ER, Mumma RO (1981) J Agric Food Chem 29:559
876. Vallejo RP, Bogus ER, Mumma RO (1982) J Agric Food Chem 30:572
877. Randall G, Jacobs P (1985) Exp Hematol 13:874
878. Wong PY, Cheung M, Yip TK, Mee AV (1986) Transplant Proc 18:780
879. Pfadenhauer EH, Jones CE, Maxwell KW (1983) J Pharm Sci 72:914
880. Thomas K, Vankrieken L, Van Lierde M (1982) Gynecol Obstet Invest 14:151
881. Horowitz PE, Spector S (1973) J Pharmacol Exp Ther 185:91
882. Young RL, Dixon WR, Mohacsi E, Liebman A, Szuna A (1976) Fed Proc 35:708
883. Dixon R, Hsiao J, Taaffe W, Hahn E, Tuttle R (1984) J Pharm Sci 73:1645

884. Berkowitz BA, Ngai SH, Hempstead J, Spector S (1975) J Pharmacol Exp Ther 195:499
885. Hahn EF, Lahita R, Kreek MJ, Duma C, Inturrisi CE (1983) J Pharm Pharmacol 35:833
886. Mu J, Faraj BA, Israili ZH, Dayton PG (1974) Life Sci 14:837
887. Shei WL, Mu JY, Cunningham RF, Israili ZH, Dayton PG (1977) Psychopharmacol 53:315
888. Miwa A, Yoshioka M, Shirahata A, Nakagawa Y, Tamura'1976) Chem Pharm Bull 24:1422
889. Brown BL (1981) J Forensic Sci 26:766
890. Cheetham RC, Fletcher SM, Whitehead PH (1983) Forensic Sci Int 22:195
891. Yamamoto Y, Tsutsumi A, Ishizu H (1984) Forensic Sci Int 24:69
892. Hampl R, Picha J, Chundela B, Stárka L (1978) J Clin Chem Clin Biochem 16:279
893. Jondorf WL, Moss MS (1978) Xenobiotica 8:197
894. Chapman DI (1979) Irish Vet J 33:37
895. Haywood PA (1980) Proceedings of the 3rd International Symposium on Equine Medicine, University of Kentucky Dept of Veterinary Science, p 273
896. Hoffmann B, Blietz C (1983) J Animal Sci 57:239
897. Morris HG, de Roche G, Caro CM (1973) Steroids 22:445
898. Johnson MW, Youssefnejadian E, Craft I (1976) J Steroid Biochem 7:795
899. Mizuchi A, Okada N, Henmi Z, Miyachi Y (1975) Steroids 26:635
900. Mizuchi A, Miyachi Y, Tamaki K, Kukita A (1976) J Invest Dermatol 67:279
901. Theodorakis MC, Stefanakou SVS (1981) Drug Develop Ind Pharm 7:411
902. Aherne GW, Littleton P, Thalen A, Marks V (1982) J Steroid Biochem 17:559
903. Nieuweboer B, Lübke K (1977) Horm Res 8:210
904. Dumasia MC, Chapman DI, Moss MS, O'Connor C (1973) Biochem J 133:401
905. Meikle AW, Lagerquist LG, Tyler FH (1973) Steroids 22:193
906. Hichens M, Hogans AF (1974) Clin Chem 20:266
907. Lee Y-S, Oshawa N (1974) Endocrinol Japon 21:481
908. English J, Chakraborty J, Marks V, Parke A (1975) Eur J Clin Pharmacol 9:239
909. Zirker DK, Kreuger GG, Meikle AW (1976) J Invest Dermatol 66:376
910. Kundu N (1974) Steroids 23:155
911. Rao PN (1974) Steroids 23:173
912. Rao PN, de la Pena A, Goldzieher JW (1974) Steroids 24:803
913. Nerenberg C, Matin SB (1981) J Pharm Sci 70:900
914. Colburn WA (1975) Steroids 25:43
915. Cornette JC, Kirton KR, Duncan GW (1971) J Clin Endocrinol Metab 33:459
916. Royer ME, Ko H, Campbell JA, Murray HC, Evans JS, Kaiser DG (1974) Steroids 23:713
917. Hiroi M, Stanczyk FZ, Goebelsmann U, Brenner PF, Lumkin ME, Midhell DR (1975) Steroids 26:373
918. Colburn WA, Buller RH (1973) Steroids 22:327
919. Meikle AW, West SC, Weed JA, Tyler FH (1975) J Clin Endocrinol Metab 40:290
920. Nygren K-G, Lindberg P, Martinsson K, Bosu WTK, Johansson EDB (1974) Contraception 9:265
921. Warren RJ, Fotherby K (1974) J Endocrinol 62:605
922. Hillier SG, Read GF (1975) J Endocrinol 67:5P
923. Morris SE, Cameron EHD (1975) J Steroid Biochem 6:1145
924. Walls C, Vose CW, Horth CE, Palmer RF (1977) J Steroid Biochem 8:167
925. Cameron EHD, Morris SE, Nieuweboer B (1974) J Endocrinol 61:39
926. Garza-Flores J, Diaz-Sanchez V, Bedolla-Tovar N, Lozano-Ruy A (1983) Steroids 41:693
927. Stanczyk FZ, Hiroi M, Goebelsmann U, Brenner PF, Lumkin ME, Mishell DR (1975) Contraception 12:279
928. Jondorf WR (1977) Xenobiotica 7:671
929. Jondorf WR, MacDougall DF (1977) Vet Rec 100:560
930. Jansen EHJM, van den Berg RH, Zomer G, Stephany RW (1985) J Clin Chem Clin Biochem 23:145
931. Sudo K, Masuoka M, Hiraga K, Yoshida K, Makayama R (1984) J Pharm Dyn 7:378
932. Shaw MA, Back DJ, Cowie AM, Orme MCLE (1985) J Steroid Biochem 22:111
933. Colburn WA, Buller RH (1973) Steroids 21:833
934. Meikle AW, Weed JA, Tyler FH (1975) J Clin Endocrinol Metab 41:717
935. Colburn WA, Sibley CR, Buller RH (1976) J Pharm Sci 65:997

936. Schalm SW, Summerskill WHJ, Go VLW (1976) Mayo Clin Proc 51:761
937. Chakraborty J, English J, Marks V, Dumasia MC, Chapman DJ (1976) Br J Clin Pharmac 3:903
938. Loo JCK, Jordan N (1978) Res Commun Chem Pathol Pharmacol 21:523
939. Olivesi A, Smith DS, White GW, Pourfarzaneh M (1983) Clin Chem 29:1358
940. Colburn WA (1974) Steroids 24:95
941. Ponec M, Frölich M, de Lijster A, Moolenaar AJ (1977) Arch Dermatol Res 259:63
942. Haack D, Vecsei P (1982) Arzneim.-Forsch/Drug Res 32:832
943. Hoffmann B, Oettel G (1976) Steroids 27:509
944. Castonguay A, Van Nunakis H (1982) Methods in Encymology 84:641
945. Cernosek SF, Langone JJ, Gjika HB, Van Vunakis H (1973) Fed Proc 32:511
946. Langone JJ, Gjika HB, Van Vunakis H (1973) Biochem 12:5025
947. Langone JJ, Van Vunakis H, Hill P (1975) Res Commun Chem Pathol Pharmacol 10:21
948. Langone JJ, Van Vunakis H (1982) Methods in Enzymology 84:628
949. Knight GJ, Wylie P, Holman MS, Haddow JE (1985) Clin Chem 31:118
950. Jones SR, Amatayakul S (1985) Clin Chem 31:1076
951. Haines CF, Mahajan DK, Miljković D, Miljković M, Vesell ES (1974) Clin Pharmacol Ther 16:1083
952. Matsushita H, Noguchi M, Tamaki E (1974) Biochem Biophys Res Commun 57:1006
953. Gehlbach SH, Perry LD, Williams WA, Freeman JI, Langone JJ, Peta LV, Van Vunakis H (1975) Lancet 1:478
954. Matsukura S, Sakamoto N, Imura H, Matsuyama H, Tamada T, Ishiguro T, Muranaka H (1975) Biochem Biophys Res Commun 64:574
955. Shen W-C, Van Vunakis H (1976) Fed Proc 35:667
956. Langone JJ, Franke J, Van Vunakis H (1974) Arch Biochem Biophys 164:536
957. Godal A, Olsnes S, Pihl A (1981) J Toxicol Environment Health 8:409
958. Langone JJ, Van Vunakis H (1976) J Nat Cancer Inst 56:591
959. Sizaret P, Malaveille C, Montesano R, Frayssinet C (1982) J Nat Cancer Inst 69:1375
960. Qian G-S, Yasei P, Yang GC (1984) Analyt Chem 56:2079
961. Fiume L, Busi C, Campadelli-Fiume G, Franceschi C (1975) Experientia 31:1233
962. Faulstich H, Trischmann H, Zobeley S (1975) Febs Lett 56:312
963. Fiume L, Busi C, Costantino D (1978) Collect Med Leg Toxicol Med 10:37
964. Busi C, Fiume L, Costantino D (1978) Amanita Toxins and Poisoning, Faulstich H, Kommerell B, Wieland T (eds), Verlag Gerhard Witzstrock, p 43
965. Faulstich H, Zobeley S, Trischmann H (1982) Toxicon 20:913
966. Andres RY, Frei W, Gautschi K, Vonderschmitt DJ (1986) Clin Chem 32:1751
967. Bauminger S, Lindner HR, Perel E, Arnon R (1969) J Endocrinol 44:567
968. Chu FS (1984) J Food Protect 47:562
969. Aalund O, Brunfeldt K, Hald B, Krogh P, Poulsen K (1975) Acta Pathol Microbiol Scand 83:390
970. Johnson HM, Frey PA, Angelotti R, Campbell JE, Lewis KH (1964) Proc Soc Exp Biol Med 117:425
971. Harvey MH, Morris BA, McMillan M, Marks V (1985) Human Toxicol 4:503
972. Coulter AR, Sutherland DK, Broad AJ (1974) J Immunol Meth 4:297
973. Mohri Z-I, Yasuda T, Kawahara K, Shono F, Yoshitake A (1985) J Pharmacobio.-Dyn 8:227
974. Bizollon CA, Rocher JP, Chevalier P (1982) Eur J Nucl Med 7:318
975. Wyse BW, Hansen RG (1976) Fed Proc 35:660
976. Wyse BW, Hansen RG (1977) Fed Proc 36:1169
977. Conrad DH, Wirtz GH (1973) Immunochem 10:273
978. Gemeiner M (1976) Mikrochim Acta 2:161
979. Fraher LJ, Adami S, Clemens TL, Jones G, O'Riordan JLH (1983) Clin Endocrinol 18:151
980. Hummer L, Nilas L, Tjellesen L, Christiansen C (1984) Scan J Lab Clin Invest 44:163
981. De Leenheer AP, Bauwens RM (1985) Clin Chem 31:142
982. Johnston MFM, Eisen HN (1974) Biochem 13:5547
983. Johnston MFM, Eisen HN (1976) J Immunol 117:1189

12 Glossary

Adjuvant:	A formulation that is blended with an immunogen prior to injection in order to enhance antibody formation.
Affinity:	Energy of binding between antigen and antibody. Often used synonymously with "avidity". Strictly, affinity refers to the antigen, avidity to the antibody.
Antibody:	A serum protein of the immunoglobulin type. There are five classes, IgA, IgD, IgE, IgG and IgM. An IgG antibody (the class used in radioimmunoassay) has two identical binding sites.
Antigen:	A substance that can be bound specifically by an antibody.
Antigenic determinant:	The part of an antigen that is bound specifically by an antibody. Synonymous with "epitope".
Ascites:	An accumulation of serous fluid in the peritoneal cavity.
Avidity:	See "affinity".
Bound fraction:	The portion of total antigen bound by antibody.
Class-specific assay:	Synonymous with "general assay".
Clone:	A group of identical cells arising from a single ancestor.
Conjugation labelling:	Preparation of a labelled compound by covalent attachment of a "prosthetic group", i.e. a structure that can be labelled with, for instance, a radioisotope, either before or after attachment.
Disequilibrium assay:	An assay in which the bound and free fractions are separated before antigen-antibody equilibrium is attained. See "late-addition assay".
Epitope:	Synonymous with "antigenic determinant".
Equilibrium assay:	An assay in which the bound and free fractions are separated after antigen-antibody equilibrium is attained.
Free fraction:	The portion of total antigen not bound by antibody.
General assay:	An assay that can detect a range of structurally related substances.
Hapten:	A substance of low molecular weight, e.g. a drug, that can be bound by an antibody but is not immunogenic unless attached to a large molecule such as a protein.
Heterogeneous assay:	An assay in which separation of the bound and free fractions is necessary.
Heterologous assay:	An assay in which dissimilar linkages attach the prosthetic group to the hapten in the radioligand and the hapten to the protein in the immunogen.
Heteroscedasticity:	Variable precision over the range of a calibration curve.
Homogeneous assay:	An assay in which physical separation of the bound and free fractions is not required.
Homologous assay:	An assay in which the linkages between the prostethic group and hapten in the radioligand and the hapten and protein in the immunogen are identical.

Hybridoma:	A hybrid cell formed by fusing a spleen cell with a myeloma cell.
Incubation:	The period allowed for antigen-antibody equilibration to occur.
Immunogen:	A substance that will induce antibody formation when injected parenterally into an animal.
Immunoreactivity:	Affinity of an antigen for an antiserum.
Label:	The radioisotope or other substance used as a marker in an immunoassay.
Late-addition assay:	One reagent, normally radioligand, added at a later stage than the others. Separation of the bound and free fractions usually effected before antigen-antibody equilibrium attained; see "disequilibrium assay".
Ligand:	Synonymous with "antigen".
Melanoma:	Melanin-pigmented tumour.
Monoclonal antibodies:	Antibodies of identical structure produced by a clone.
Parenteral:	Not via the alimentary canal.
Polyclonal antiserum:	Serum from an immunised animal containing antibodies against the immunogen of varying avidity and specificity.
Prosthetic group:	See "conjugation labelling".
Radioiodination:	Incorporation of a radioactive isotope of iodine, usually ^{125}I, into a molecule.
Radioligand:	An antigen labelled with a radioactive isotope.
Specific activity:	Amount of radioactivity per unit mass.
Specific assay:	An assay designed to detect one compound only.
Specificity:	The degree to which an antibody or antiserum binds a single compound in preference to structurally related compounds.
Titre:	The dilution of antiserum that binds 50% of an arbitrary amount of radioligand.
Total tubes:	Assay tubes containing radioligand only.
Tracer:	Synonymous with "label".

Developments in Fingerprint Visualisation

C. A. Pounds
Central Research Establishment, Home Office Forensic Science Service, Aldermaston, Reading, Berkshire, RG7 4PN, United Kingdom

There are a variety of methods for visualising latent and contaminated fingerprints drawing on a wide area of science, from surface physics to chemistry. Although simple and effective methods, such as ninhydrin and powders, will remain the routine techniques for fingerprint technicians, there are now many methods available which supplement these development techniques, some of which are complicated and/or expensive. Valuable methods now available to us include cyanoacrylate ester vapour which is a simple easy to use technique and high powered light sources, such as argon ion lasers which although expensive, have increased the number of fluorescence techniques available. Recently, research into biological and bacteriological methods has offered potential for the future.

This article reviews contributions, taken substantially from the open literature, which demonstrate the diverse knowledge now needed by those interested in fingerprint development techniques. It is intended to give scientists working in the field of fingerprint detection a thorough, up to date, picture of research in both physical and chemical techniques.

A general introductory section describes the make up of a latent fingerprint from material secreted by the various types of gland in the body. This is followed by discussing individual methods of detection. After a short introduction to a method among areas discussed are, mode of action, development conditions, advantages and drawbacks, suitable surfaces, various formulations and procedures for enhancement. Finally in some instances, suggestions for future research are made.

1 General Introduction to Latent Fingerprints 93

1.1 Introduction 93
1.1.1 Sweat Gland Secretions 93

2 Methods of Detection 95

2.1 Ninhydrin 95
2.1.1 Reagent Formulation 95
2.1.2 Development Conditions 97
2.1.3 Enhancement Procedures 97
2.1.4 Ninhydrin Analogues 99
2.2 Luminescence 100
2.2.1 Luminescence of Natural or Contaminated Components of Fingerprints 100
2.2.2 Derivatisation Reagents 101
2.2.3 Vapour Phase Reagents 103
2.3 Cyanoacrylate Ester (Super Glues) 104
2.3.1 Enhancement Procedures 105
2.3.2 Interference with Firearms Examination 106
2.4 Surfactant Controlled Reagents 106

Forensic Science Progress Vol. 3

2.4.1 Physical Developer 106
2.4.2 Small Particle Reagent 107
2.5 Iodine 108
2.5.1 Iodine Fuming Methods 108
2.5.2 Fixing of Iodine Fumed Fingerprints 109
2.5.3 Use of Iodine Fuming to Reveal Fingerprints on Skin 109
2.6 Dyes and Stains 109
2.6.1 Development of Fingerprints on Adhesive Tapes 109
2.6.2 Enhancement of Blood Fingerprints 110
2.7 Silver Reagents 111
2.7.1 Silver Nitrate 111
2.7.2 Other Silver Reagents 111
2.8 Powders 112
2.8.1 Fluorescent Powders 113
2.9 Metal Deposition (MD) 113
2.10 Radioactive Techniques 114
2.11 Electronography 114
2.12 Dimethylaminocinnamaldehyde (DMAC) 115
2.13 Biological Detection Methods 115
2.13.1 Antisera and Lectins 115
2.13.2 Bacteriological Techniques 115

References 116

1 General Introduction to Latent Fingerprints

1.1 Introduction

The comparison of fingerprints is a well established way of identifying an individual, whether he/she is a disaster victim or a criminal who leaves marks at the scene of a crime. Recent research in DNA 'Fingerprinting' of body fluids may, however, rival the technique for discrimination purposes [1]in the future.

Characteristic patterns on each finger are formed during the development of the foetus and, provided no serious injuriy occurs, they remain unchanged throughout life. Details of identification and classification of such patterns are outside the scope of this review but may be found in texts by Olsen [2], Moenssens [3] and Cowger [4].

There are three ways by which a fingerprint may be left on an object:
(i) As a visible mark made from contamination on the finger with, for example, blood, paint, oil, grease, etc.
(ii) As an impression in a soft surface such as putty.
or (iii) As a deposition of sweat.

Fingerprints found at scenes of crime or on feloniously handled material generally arise from sweat and, to a lesser extent, blood or other contaminants.

1.1.1 Sweat Gland Secretions

There are three types of gland in the body the secretions of which contribute to a fingerprint deposit viz, eccrine, apocrine and sebaceous glands. A review on the morphological and cytological features of the sweat glands has been published by Weiner and Hellmann [5].

Eccrine glands are widely distributed throughout the body and, importantly, are particularly numerous on the palms of hands and the soles of feet. Chemicals secreted by these glands result from material in the bloodstream crossing the gland barrier [6]. Apart from its water content, eccrine sweat contains up to 1% of other substances the main constituents of which are shown in Table 1.

Table 1. Major constituents of eccrine sweat

Inorganic constituents	Organic constituents
Chlorides	Amino Acids
Metal ions	Urea
Ammonia	Lactic acid
Sulphate	Sugars
Phosphate	Creatinine
	Choline
	Uric acid

(Data taken from Knowles [11])

Sebaceous glands are associated with hair follicles and are mainly located on the forehead, chest back and abdomen but are absent on palms, or soles and the dorsal surface of the feet[7]. Lipid material from the sebaceous glands is, however, transferred to fingers by touching the face or head. Table 2 lists secretions of the sebaceous gland.

Table 2. Constituents of serum

Compound	Relative amount present 12	13
Glycerides (Mono-di-and tri)	31.7%	35–60%
Free fatty acids	29.6	15–30
Wax esters	20.2	12–16
Squalene	12.8	10–12
Sterol esters	3.3	1–3
Sterols	2.4	1–3
Hydrocarbons	0.8	1–3

(Data taken from Lewis and Hayward[12] and Powe[13]. Sebum synthesized in the sebaceous gland contains little if any, fetty acids[14], which are formed from the hydrolysis of triglycerides by bacterial lipases[15].

Apocrine glands, which usually contribute little to the fingerprint deposit, are found in the anogenital regions and mammary areolea but the vast majority are concentrated in the axillary regions of the body. Table 3 lists secretions of these glands.

Table 3. Secretions from the apocrine gland

Inorganic constituents	Organic constituents
Iron	Proteins
	Carbohydrates
	Cholesterol

(Data from Knowles[11])

The amount of material discharged from the glands varies between individuals and is influenced by mental, sensory, thermal, emotional and psychic stimuli[8].

To exploit the differences occurring in the various secretions from glands a range of methods needs to be available which can be used sequentially, each one reacting with different constituents of the fingerprint deposit. Series of this type have been suggested by Goode and Morris[9] and a comprehensive manual of recommended fingerprint detection techniques, and sequential processing routes for all surfaces likely to be encountered in casework, has been published recently by the UK Home Office[10].

2 Methods of Detection

2.1 Ninhydrin

The most common use of ninhydrin is as a colourimetric reagent for the detection of amino acids in thin layer and paper chromatography. Ruhemann [16] is credited with having first prepared the reagent in 1910 and outlined the synthesis of ninhydrin [17], proposed its structure to be 2,2–dihydroxy–1,3–indanedione and described the mechanism of reaction with amino acids, the product of which is referred to as 'Ruhemann's Purple'. The mechanism is however far more complex than first thought and a recent review by Bottom et al. [18] summarises the current knowledge (Fig. 1).

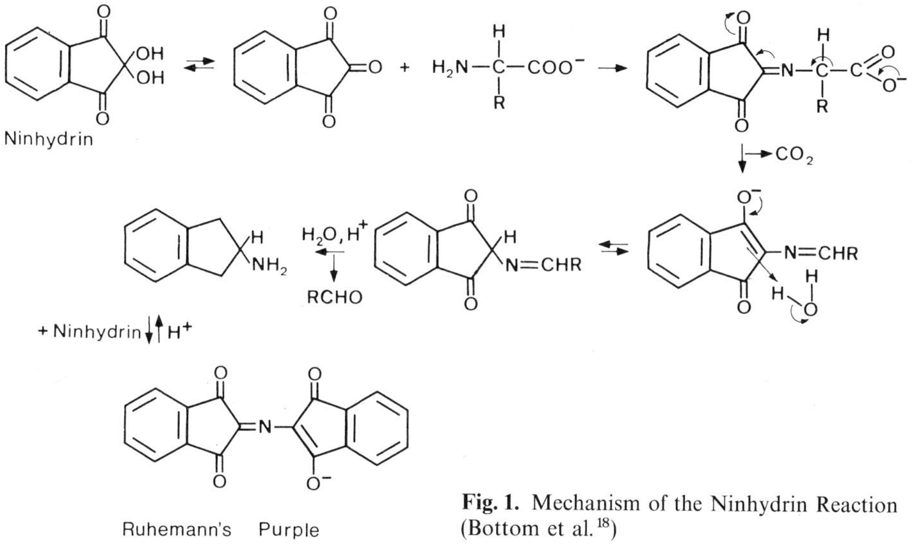

Fig. 1. Mechanism of the Ninhydrin Reaction (Bottom et al. [18])

The use of ninhydrin for the detection of latent fingerprints was first suggested by Oden and von Hofsten [19] who in 1954, noted that purple coloured fingerprints were revealed on paper chromatograms that had been handled. Oden [20] was granted a British Patent for the use of ninhydrin as a fingerprint reagent in 1957. Since then ninhydrin has been commonly employed for revealing latent fingerprints on paper and porous surfaces where it reacts with the amino acid fraction of fingerprint deposits which originate from the eccrine swear glands.
Table 4 summarises many of the formulations and development conditions described in the literature.

2.1.1 Reagent Formulation

Oden and von Hofsten [19] originally detected latent fingerprints on paper by applying a solution of ninhydrin in acetone, but this procedure was later modified by Oden [20] who reported the addition of acetic acid to the formulation. The work of Moore and Stein [21]

Table 4. Summary of ninhydrin formulations and development conditions

Authors	Formulation		Development
	% Ninhydrin	Solvent	
Oden and von Hofsten [19]	0.2	Acetone	80 °C for few minutes
Oden [20]	0.2	Acetone or ether + 4.0% acetic acid	80 °C to 120 °C for 1–3 minutes
Crown [25]	0.5 0.5 0.75	Acetone Ethyl ether Petroleum ether + 4% methanol	Air curing
Fritz and Jordan [44]	1.0	Acetone + 6% water + 0.3% acetic acid	30 minutes UV irradiation then treatment with ninhydrin
Rispling [23]	0.7–0.8	Actone + 4% glacial acetic acid	90 °C for 3 minutes. If humidity less then 40%, steam over kettle for few seconds then heat
Linde [24]	1.0	Acetone + 3% water	60 °C in atmosphere saturated with water for 30 minutes
Morris and Goode [9]	0.5	2% Ethyl alcohol +1% acetic acid in 112-trifluoro 122-trichloroethane	60 °C for 20 minutes or 120 °C for 2 minutes
Menzel et al. [38]	Staturated solution in methanol diluted 1:5 with 112-trifluoro 122-trichloroethane		Ambient conditions for 1 day
Tighe [27]	0.5	1% Methanol + 9% ethyl acetate in 112-trifluoro-122-trichloroethane	Not given

and Lamothe and McCormick [22] showed that this was an important step as the formation of the Ruhemann's purple was found to be pH dependent. The successful use on operational material was later reported by Rispling [23] who employed a solution of ninhydrin in acetone acidified with glacial acetic acid. Linde [24] subsequently compared 16 ninhydrin formulations, including 4 acidified solutions and 5 commercially available sprays, and reported the optimum formulation to be an acetone solution with water added.

The disadvantage of all of these, when used to reveal fingerprints on documents, is that they cause inks to run. To overcome this problem, Crown [25] prepared a methanolic solution of ninhydrin which he diluted with petroleum ether. Following from this work and using non-polar solvents, Morris and Goode [26] recommended diluting an acidified ethanolic solution of ninhydrin with 1,1,2–trifluoro 1,2,2–trichloroethane (Fluorisol or Arklone P) which, in addition to being free from problems of ink stability, is non–toxic and non–flammable. This formulation was referred to as 'non–flammable ninhydrin' (NFN). A further advantage found by Morris and Goode [9] was that NFN, as well as its essential safety features, produced considerably less background colouration than other ninhydrin formulations. They further reported that their formulation, which now includes molecular sieve to absorb water, had been confirmed by UK Police Forces as a successful reagent.

Menzel [10] describes preparing the reagent by diluting a saturated methanolic solution of ninhydrin with Fluorisol, but without the addition of acid thought by many previous workers to be essential for optimum colour intensity. Using this method fingerprints were reported to be typically revealed in a few hours at ambient conditions. In a recent report Tighe [27] formulated the reagent in a solution of methanol and ethyl acetate diluted with Fluorisol. Time taken in preparation was said to be considerably reduced compared to that for the method of Morris and Goode [9].

2.1.2 Development Conditions

In the original work by Oden and von Hofsten [19] ninhydrin treatment was followed by heating samples at 80 °C for several minutes to accelerate the otherwise slow reaction. They noted that not only did the intensity of the pink coloured fingerprint increase with time but also that new fingerprints appeared as the developing process continued. Crown [25] conducting controlled tests, compared the use of a pressing iron, infra–red lamp and air curing but all showed similar colour development of fingerprints after 1 week irrespective of treatment. The air curing method tended to give less background than the heating methods and was preferred if there were no time limitations. Crown notes that ninhydrin revealed fingerprints are not permanent, with fading occurring about 1 month after optimum development.

Rispling [23] also heated samples after ninhydrin treatment but if development took place at low humidities (less than 40%) the paper required moistening with water vapour. On the basis of these findings a special laboratory was created where an atmosphere of 70–80% relative humidity (RH) was maintained. The importance of high humidity was also observed by Linde [24] who, after reviewing 16 methods of development, proposed that samples should be heated in an atmosphere saturated with water. More recently, Goode and Morris [9], following a survey of many heating and steaming procedures, recommended storing treated samples in the dark for 24 hours to 48 hours without heating. If rapid processing was required they recommended heating samples at 80 °C for 3 minutes. The workers also found that domestic steam irons, which combined both heating and steaming steps [28] could not be recommended as control of background intensity was difficult.

2.1.3 Enhancement Procedures

Herod and Menzel [29] have demonstrated that fingerprints weakly revealed by ninhydrin luminesce when irradiated with light. Red fluorescence was observed under light from an argon ion dye laser at a wavelength of 580 nm. Well developed fingerprints did not exhibit this fluorescence and it was therefore speculated that a reaction by–product or intermediate rather than Ruhemann's Purple was fluorescing.

German [30] and Creer [31] have both reported operational experience using argon ion lasers to irradiate samples treated with ninhydrin. In these cases, however, the ninhydrin developed fingerprints absorbed light while the surface backgrounds fluoresced.

In order to facilitate photography, where contrast between background and the developed fingerprint is poor, Goode and Morris [9] suggested changing the purple colour of fingerprints developed by ninhydrin. Treatment of ninhydrin developed fingerprints with a solution of zinc chloride changes the colour to orange whereas nickel chloride solution produces an orange–red colour.

The effect of complexing metals to organic structures, such as Ruhemann's Purple, often produces fluorescent species [32]. For instance, the zinc complex with ninhydrin developed fingerprints has been shown by Herod and Menzel [33] to be highly fluorescent when irradiated at a wavelength of 488nm with light produced by an argon ion laser. Menzel [34] and Creer [31] have both reported using this technique with success on operational samples. About half the cases examined by Creer, which had previously been treated with ninhydrin, produced identifiable fingerprints by fluorescence of the zinc complex.

Kobus et al. [35] have developed the technique further and have shown that fluorescence of a fingerprint, weakly developed after treatment with ninhydrin and zinc chloride, was increased when cooled to $-196\,^{\circ}C$ in liquid nitrogen. At this temperature a high intensity laser was not necessary to produce fluorescence because comparable results were obtained with a suitably filtered xenon arc lamp. As a result of their findings Kobus et al. [36] have subsequently designed a compact xenon arc lamp with incorporated filters. It was found that in dry conditions, ($<40\%$ RH) it was essential for maximum fingerprint development to steam the sample.

Recent work by Warrener et al. [37] suggests that the cadmium complex formed with Ruhemann's Purple has a number of benefits over the zinc analogue. The emission maximum was determined to be 575 nm (compared to 545 nm for the zinc complex) and this gave improved contrast on papers that fluoresce. Following from this, Stoilovich et al. [37A] investigated the photoluminescence properties of complexes produced by Ruhemann's Purple and a range of metal salts. Although many exhibited photoluminescence at liquid nitrogen temperatures they concluded that only group 2b of the periodic table were suitable for enhancing ninhydrin developed fingerprints. Cadmium nitrate was found to be less affected by extremes of heat and humidity and was therefore preferred to zinc chloride and mercuric bromide; the latter was thought usefull if a red shift is necessary (λex 525 nm; λem 600 nm). Caution must be exercised when using this technique as eadmium and mercury salts are very toxic.

An original approach to enhancement has been reported recently by Menzel et al. [38] who investigated the use of enzymes as a pre–treatment to ninhydrin. In this method, the enzymes trypsin and pronase are dusted on latent fingerprints, incubated for up to 24 hours, then treated with ninhydrin. For fresh fingerprints, up to 2 days old, a pronounced enhancement over normal ninhydrin developed fingerprints was observed for the enzyme treated fingerprints. Enhancement of older prints was, however, marginal. Although the mechanism of the enzyme action is not certain, the authors from a pragmatic standpoint thought it unimportant whether the enzymes acted as dusting powders or produced amino acids by hydrolysis of proteins present in fingerprints. However, no consideration was given to the potential hazard arising from the inhalation of enzyme dust.

2.1.4 Ninhydrin Analogues

Modifications to the ninhydrin molecule, have been undertaken by Almog et al. [39] in an attempt to produce reaction products darker coloured than Ruhemann's Purple. Syntheses based on the methods of Becker [40], Jones and Wife [41] and Meier and Lotter [42] were used to prepare ring fused and substituted ninhydrins. (Fig. 2 shows

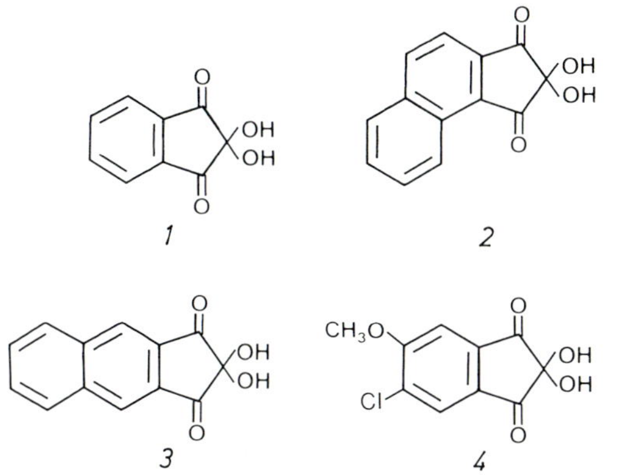

Fig. 2. Ninhydrin analogues prepared by Almog et al. [39].
1. 2,2–dihydroxy–1,3–indanedione (ninhydrin);
2. 2,2–dihydroxybenz[e]indane–1,3–dione(benzo[e]ninhydrin);
3. 2,2–dihydroxybenz[f]indane–1,3–dione(benzo[f]ninhydrin);
4. 2,2–dihydroxy–5–chloro–6–methoxyindane–1,3–dione

structures of these molecules). All three analogues prepared were found to be as sensitive as ninhydrin for revealing amino acids, but benzo(f)ninhydrin showed greater promise producing dark green fingerprints, and thereby increasing contrast with many coloured surfaces. Recently, Menzel and Almog [43] have studied the fluorescent properties of the fingerprints developed by such analogues when complexed with zinc chloride. It was found that only the benzo(f)ninhydrin complex fluoresced as intensely as the ninhydrin complex. However, because the absorption maximum of the compound was 530 nm, illumination from an argon ion laser at 488 nm or 514 nm is not effective and it means that the use of a Neodymium Yttrium Aluminium Garnet (NdYAG) laser, which emits at 532 mn, was required. They plan further studies on vicinal triketones which are not ninhydrin analogues.

For the future, the approach of Almog et al. [39] in preparing ninhydrin analogues may well lead to: —

(i) An increase in rate of reaction with amino acids.

(ii) The ability to select an analogue which produces maximum contrast with a given surface.

(iii) The potential of improved detection sensitivity by enhancing fluorescence of metal complexes.

Lennard et al. [43A] have recently synthesized methoxy, chloro, bromo, benzo and perinaphtho derivatives of ninhydrin. All developed fingerprints with the same sensitivity as ninhydrin with colours ranging from pink to dark green; photoluminescence complexes were formed on addition of zinc and cadmium salts. Preliminary results showed that some analogues may be useful when background luminescence would cause problems with ninhydrin developed fingerprints. A full evaluation will be reported later.

2.2 Luminescence

High intensity light sources, particularly lasers, have been commonly employed to induce fluorescence in natural or contaminated fingerprint deposits or where fluorescent derivatives have been prepared. The basic principles and applications have been well reviewed by Menzel [45]. In a wide ranging paper the same author [46] extended these studies by comparing argon ion, copper and NdYAG lasers for revealing fluorescent fingerprints following various processes and treatments.

2.2.1 Luminescence of Natural or Contaminated Components of Fingerprints

The visualisation of latent fingerprints by luminescence was first reported by Dalrymple et al. [47] in 1977, when fingerprints deposited on paper were illuminated by light from an argon ion laser. Using an argon ion laser, excellent ridge detail was observed for the majority of deposited fingerprints with only one fifth showing faint luminescence. When sebum rich fingerprints were deposited all could be visualised. This observation, combined with control experiments, led the authors to conclude that inherent components rather than contaminants were fluorescing. Lamps were less successful than the argon ion laser but it was thought that they had potential application for field work. The research also showed that the emission spectra of fingerprint material dissolved in methanol showed three maxima but the luminescent species were not identified. It was, however, noted that old prints gave an orange fluorescence compared to yellowish orange observed for fresh samples. Following on from this observation Duff and Menzel [48] investigated the use of luminescent components as a possible guide to determining age of a fingerprint. The luminescent compounds, however, included vitamins and not, as hoped, species whose stability was time dependent. Dikshitulu et al. [48A] have also investigated components of fingerprint deposits over periods of time in an attempt to determine age of fingerprints. Fingerprint residues from eight subjects were separated using thin layer chromatography and high pressure liquid chromatography, variation in three major and two or three minor components were observed but no useful conclusions can be drawn. Contrary conclusions from those of Dalrymple and his co–workers [47] were drawn by Salares et al. [49] who carried out two surveys of untreated, natural fingerprints using an argon ion laser for irradiation. Only one fifth of the deposited fingerprint exhibited fluorescence and they failed to observe luminescence from unstained fingers or fingerprints. Salares et al. suggested that is was contamination of the fingerprint deposit which was producing the fluorescence.

Dalrymple [50] has reported success in using narrow band filters when both fingerprint and background fluoresce. This technique was illustrated in two cases where interference from background luminescence was significantly diminished. For weakly fluorescing fingerprints, which may take hours to record by photography, Matharu et al. [51] designed a computer controlled scanner to aid the movement of the laser beam during photography.

The use of luminescent techniques in casework has been studied in particular by Creer [31] who reports considerable success when a 2 W argon ion laser was employed to reveal fluorescent fingerprints on 121 out of 396 exhibits considered unsuitable for conventional treatments. Tests with a 12 W argon ion laser revealed fingerprints that may have been missed with a lower powered laser. No attempt was made to identify the source of fluorescence.

2.2.2 Derivatisation Reagents

For a reagent to be successful in producing fluorescent products with constituents of fingerprint deposits there are three requirements: —

a) The reagent should preferably not be fluorescent.

b) The fluorescence of the final product should not correspond with any background fluorescence e.g. optical brighteners in paper [52].

c) It is preferable for the reagent to be specific for a particular fingerprint component.

Reagents which fulfill at least some of these requirements are reviewed in the following sections and also by Seitz [32].

7–chloro–4–nitrobenzo–2–oxa–1,3–diazole (NBD–Cl)

Ghosh and Whitehouse [53] first reported on the use of NBD–Cl as a fluorogenic reagent for amino acids and other amines (Fig. 3). Its use for visualising latent fingerprints was later reported by Salares et al. [49] on untreated fingerprints which did not fluoresce under argon ion laser illumination. After treatment with the reagent, fluorescent fingerprints were observed with the technique showing improved sensitivity over ninhydrin. The work of Salares et al. was supplemented by Warrener et al. [54] who used the reagent to visualise a large number of luminescent fingerprints on papers. These workers claimed that use of a 3W argon ion laser had little advantage over a suitably filtered xenon arc lamp. A major disadvantage of this derivatisation technique, however, was non–specific reaction of the reagent with some constituents of paper. This caused emissions similar to those from the derivatised fingerprint.

The use of several irradiation sources for visualising NDB–Cl/amino acid fluorescence of latent fingerprints has been reported by Creer [31] using portable sources such as a xenon arc lamp, tungsten halogen lamp and a 25 mW laser. Apart from noting that the laser gave only faintly visible fluorescence, no detailed comparisons of the

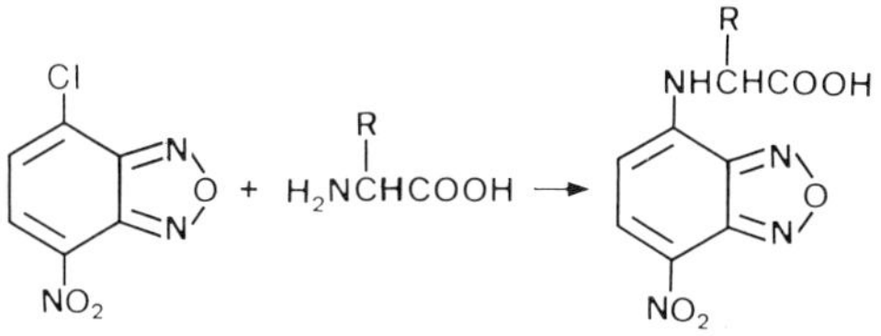

Fig. 3. Reaction of NBD–Cl with amino acids

effectiveness of the light sources were made. A detailed study of this reagent has been made by Stoilovic et al. [28] who, using a xenon arc lamp, compared the effectiveness of the reagent with ninhydrin for fingerprints deposited on paper. For fresh (up to a month old) or old (17 months old) fingerprints there was little difference between the two techniques. For fingerprints of intermediate ages (3 to 9 months old), NBD–Cl treatment showed superior sensitivity over ninhydrin.

It should be noted, however, that NBD–Cl is now suspected of being carcinogenic [55, 56].

1–dimethylaminonaphthalene–5–sulphonyl chloride (Dansyl Chloride)

Dansyl chloride was first used by Weber [57] to prepare fluorescent conjugates of albumin. Its reaction with amino acids to produce fluorescent derivatives has been reported by Gray and Hartley [58] and reviewed by Seiler [59] (Fig. 4). The extension of use

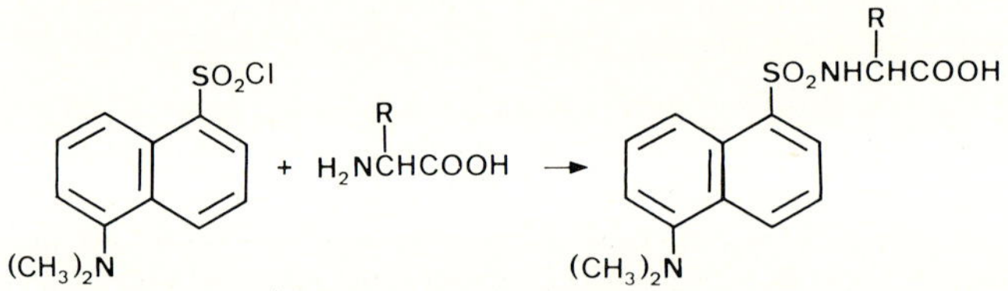

SO_2Cl, $(CH_3)_2N$ + $H_2N\overset{R}{C}HCOOH$ → $SO_2NH\overset{R}{C}HCOOH$, $(CH_3)_2N$

Fig. 4. Reaction of dansyl chloride with amino acids

of this reagent to reveal latent fingerprints has been suggested recently by Burt and Menzel [60] for paper surfaces that fluoresce strongly under light from an argon ion laser. These workers also carried out a brief comparison of the reagent with ninhydrin for revealing latent fingerprints on dark or coloured paper and found dansyl chloride to be superior.

4–phenylspiro(furan–2–(3H), 1′–phthalan)–3,3′–dione (Fluorescamine)

The use of fluorescamine to produce fluorescent products with amino acids was first reported by Udenfriend et al. [61] in 1972 (Fig. 5). Later work by Felix and Jimenez [62] and Touchstone et al. [63] reported on the stability of the fluorescent product and the time dependence of the fluorescence. The first use of the reagent for revealing latent fingerprints was reported by Olsen [2] who used a formulation similar to that described by Felix and Jimenez [62] i.e. solutions of fluorescamine in acetone with the addition of

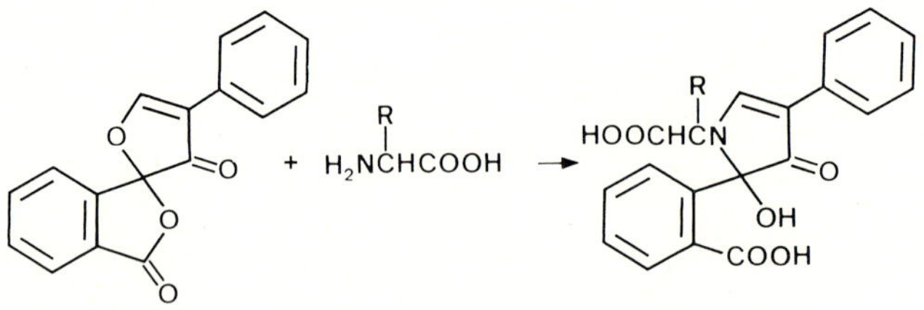

O, =O, O, O + $H_2N\overset{R}{C}HCOOH$ → $HOOC\overset{R}{C}HN$, =O, OH, COOH

Fig. 5. Reaction of fluorescamine with amino acids

triethylamine to stabilise the fluorescent product. Menzel [45] shows that treated fingerprints can fluoresce under irradiation from an ultra violet light source. Apart from these papers there have been no reports of the successful use of this reagent to reveal latent fingerprints.

O–phthalaldehyde

The reaction of amino acids with o–phthalaldehyde to give a fluorescent product was first reported by Roth [64] using an alkaline solution of the compound in the presence of a reducing agent such as 2–mercaptoethanol (Fig. 6). Simons and Johnson [65] later

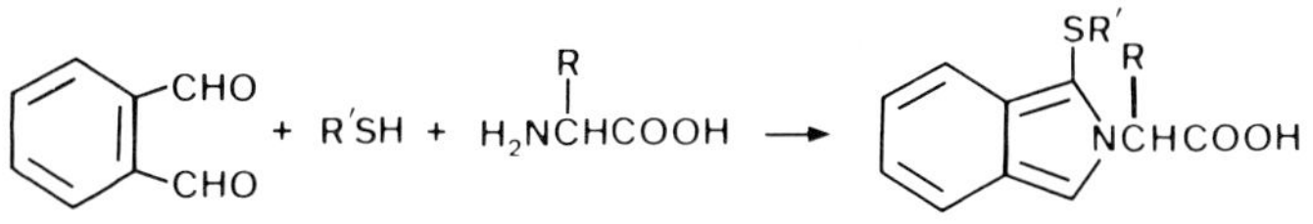

Fig. 6. Reaction of o–Phthalaldehyde with amino acids

found that 2–mercaptoethanol could be successfully substituted by ethanethiol. The sensitivity of the reagent compared to that of fluorescamine was studied by Benson and Hare [66] who demonstrated the amino acid derivative to be five times more intense than the fluorescamine product. The use of the reagent to develop latent fingerprints has been reported by Mayer et al. [67] employing a buffered solution of the reagent to which was added a detergent and 2–mercaptoethanol. Fluorescence was observed under irradiation from an ultra violet lamp.

A problem with the use of this reagent is that emission from optical brighteners present in paper [52] may interfere with the fluorescence produced by the treated fingerprint. This problem might be overcome by the synthesis of derivatives of o–phthaladehyde which should have different excitation and emission spectra from those of optical brighteners.

2.2.3 Vapour Phase Reagents

Several attempts have been made to coat latent fingerprints by exposure to vapours from fluorescent components. In a study by Almog and Gabay [68] samples deposited on paper, were exposed to fumes from a range of heated fluorescent compounds. When viewed under ultra violet irradiation anthranilic acid fuming gave best results for fresh prints (up to 1 day old) while for prints several days old, the best results were obtained from treatment with anthracene. It was also reported that antimony trichloride gave impressive fluorescent fingerprints.

Vapour phase reagents have also been used in an attempt to solve the difficult problem of revealing fingerprints on skin. Menzel [69] has reported that rhodamine 6G, when evaporated onto latent fingerprints placed on dead or live skin, produced fluorescence on irradiation with an argon ion laser. In the case of dead skin, prints could be observed one day after deposition but for older prints the success rate decreased considerably. Fingerprint visualisation on live skin was, however, poor. No success on operational cases has been reported when this technique was used on skin.

Other Reagents

In an attempt to make blood fingerprints fluorescent, several reagents were investigated by Fischer [70]. The most successful method employed formic acid and hydrogen peroxide in absolute alcohol. Although this treatment produces an intense orange fluorescence on irradiation with light from an argon ion laser, no success in case samples has been reported.

Future Work

Lasers commonly employed in fingerprint detection, are expensive, need water cooling and special power supplies. For the future, cheap, portable, high intensity, variable wavelength light sources are required. There are indications of developments in this respect. The xenon lamp reported by Warrener et al. [54] has some portability as has a similar device described by Kent [54A]. Watkin and Carey have discussed the modification of a commercially available indium arc source [54B], whilst more recently Creer [54C] has found a portable commercial 200 mW argon ion laser and light guide to be effective at crime scenes. Additionally, knowledge of fluorescent species present in fingerprint deposits, whether natural or from contamination, is scant. More information would enable irradiation wavelengths to be matched to absorption maxima of fluorescent species, thus improving sensitivity.

A further unresolved problem is that there is no derivatization reagent that is really effective in producing a fluorescent species with components of fingerprint deposit. As methods involving fluorescence are potentially much more sensitive than normal visual methods, fluorescent derivatives of amino acids are desirable.

2.3 Cyanoacrylate Esters (Super Glues)

The use of cyanoacrylate esters as adhesives is well established for industrial and domestic use. Surfaces are bonded together on polymerisation of the cyanoacrylate ester when water vapour, present in the atomsphere, acts on the acidic stabiliser present in the glue. The mechanism has been outlined by Lee and Gaensslen [71] (Fig. 7) in a paper describing the use of cyanoacrylate ester vapour to reveal latent fingerprints. The reaction product is a white polymer which deposits along the ridges of a fingerprint. The most common constituents of super glues are methyl or ethyl

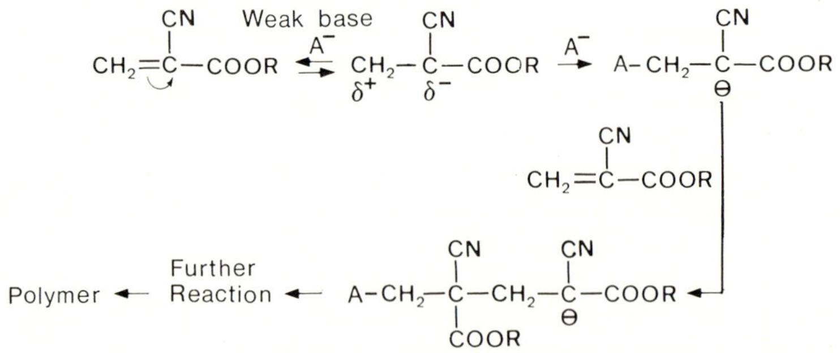

$$CH_2{=}C(CN){-}COOR \xrightleftharpoons[]{\text{Weak base } A^-} \overset{\delta^+}{CH_2}{-}\overset{\delta^-}{C}(CN){-}COOR \xrightarrow{A^-} A{-}CH_2{-}C^{\ominus}(CN){-}COOR$$

$$A{-}CH_2{-}C^{\ominus}(CN){-}COOR + CH_2{=}C(CN){-}COOR \longrightarrow A{-}CH_2{-}C(CN)(COOR){-}CH_2{-}C^{\ominus}(CN){-}COOR \xrightarrow{\text{Further Reaction}} \text{Polymer}$$

Fig. 7. Mechanism of polymerisation of cyanoacrylate esters (Lee and Gaensslen [71])

cyanoacrylate esters but as the former is more toxic than the higher homologue, the ethyl ester is most likely to be the major constituent [72].

Reed [73] reported in 1980, that latent fingerprints could be successfully revealed on a wide range of surfaces e.g. polythene, glass, etc, although details of the method were not given. By exposing exhibits to the ester vapour for several hours or up to several days in a glass tank Kendall [74] was able to reveal fingerprints on a wide range of surfaces. In later work Kendall and Rehn [75] suggested that development times could be shortened to about 1 hour by exposing exhibits to the dense white fumes produced by adding cyanoacrylate ester to dried cotton pads which had previously been impregnated with sodium hydroxide. Anhydrous sodium carbonate used in a similar manner has been reported by Martindale [76]. An alternative method of producing high concentrations of vapour was achieved by Olnik [77] by heating the ester in an aluminium boat over an electric light bulb. The method was used effectively on plastic bags, rubber gloves, electrical tape, galvanised metal, guns and leather. The presence of aluminium was not only a convenient way of containing the ester but was also thought to act as a retardant to rapid polymerisation. Methods designed to accelerate fuming were made by Besonen [78] and Woods [79] who heated the ester over a 1000 watt light bulb. The optimum temperature for heating the ester has been reported recently by Mock [80] to be between 90 and 100 °C. The author, however, added a note of caution when using uncontrolled heating, as above 205 °C hydrogen cyanide is released.

Kendall [74] has noted that humidity is a factor in cyanoacrylate development; Sahs and Wojcik [81] have further suggested from experiments undertaken by them that freezing of the sample in a refrigerator followed by thawing is particularly effective in assisting development of latent fingerprints.

A commercial source of the reagent in which the ester is incorporated into a gel in a reusable pouch, called "Hard Evidence", has been developed by Loctite Ltd. A similar approach has been recently reported by Gilman et al. [82] who sandwiched a mixture of ester and petroleum jelly between acetate sheets. The sheets are pulled apart and fuming commences, after the fuming treatment has been completed the sheets are resandwiched for further use.

2.3.1 Enhancement Procedures

Contrast between a fingerprint developed by cyanoacrylate ester and the host surface is sometimes poor. For such circumstances Woods [79] used a mixture of magnetic powder and standard fingerprint powder to dust the fumed fingerprint. Dusting procedures have been developed further by Illsley [83] who successfully used a fingerprint powder for fumed fingerprints on knives.

Enhancement of fingerprints developed by cyanoacrylate esters has also been achieved by the use of fluorescent dyes. For weakly developed fingerprints on polythene bags, Kerr et al. [84] formulated dusting powders incorporating fluorescein. A more extensive study by Kobus et al. [85] examined a selection of such dyes on various surfaces. The authors suggested treating weakly revealed fingerprints on clear or white polythene with gentian violet, those on aluminium with coumarin 540 and dusting with black fingerprint powder for white polythene. A xenon arc lamp was used successfully as an excitation source for the fluorescent dye coumarin 540. Treatment with this dye was found to be effective even when no cyanoacrylate fumed fingerprint was discernible. This same approach has also been followed by Menzel et al.[86] who reported the use of solution staining with rhodamine 6G, ninhydrin followed by zinc chloride, or dusting with magnetic powders blended with rhodamine 6G. An argon ion laser was used to excite fluorescence. In circumstances where delicate handling of exhibits such as audio/visual cassettes or glazed papers is required, Vaughn[87] has suggested treating the cyanoacrylate fumed fingerprint with vapour from heated rhodamine 6G followed by laser illumination.

Further sources of difficulty are surfaces which themselves fluoresce strongly e.g. adhesive tapes. For such cases Menzel[88] suggests using 3,3′–diethyloxadicarbocyanine iodide (DODC) which fluoresces under light emitted by a NdYAG laser at 532 nm. Alternative fuming techniques were developed by Burt and Menzel[60] who combined cyanoacrylate ester fuming and treatment with either solutions containing, or vapours of, 9–methylanthracene, coumarin 535 or rhodamine 6G. An argon ion laser and a UV lamp were used to irradiate the fluorescent dyes.

2.3.2 Interference with Firearms Examination

The use of cyanoacrylate esters to reveal fingerprints on weapons has caused concern that the white deposits may affect firearm examinations. Arnold and Gallant[89] examined spent bullets and cartridge cases before and after cyanoacrylate fuming for rifling, breech, extractor, ejector or firing pin marks. All were identified to each firearm without difficulty after the white deposit was removed with acetone. There is, however, no report on the effect which the procedure has on other aspects of firearm examination such as trigger pressure measurement.

Future Work

The method of cyanoacrylate ester or super glue fuming has rapidly gained popularity as an effective method for a wide range of surfaces. However, from the variety of methods and conditions of use in the literature there is an obvious need to optimise this process, including the temperature and humidity conditions of vapour evolution, the humidity conditions for treatment and the subsequent use of fluorescers. Assessments against other techniques would enable the examiner to select objectively whether cyanoacrylate ester fuming is the most appropriate method.

2.4 Surfactant Controlled Reagents

2.4.1 Physical Developer

Physical development is a photographic process which works by depositing nascent silver, formed in the developer, onto nuclei of a latent image. A general survey and photographic principles have been published by Jonker et al.[90]. Classical wet physical developers[91] were found to be capable of revealing latent fingerprints but instability barred their routine use[9].

The major breakthrough in the technique arose from Jonker et al.[92], working in the field of photofabrication of printed circuits who in 1969 reported a stabilized system based on silver and a ferrous/ferric redox couple. This later proved to be successful in revealing latent fingerprints[9].

The generalised reversible reaction taking place can be expressed by the reaction: —

$$Fe^{2+} + Ag^{+} \leftrightharpoons Ag + Fe^{3+}$$

Stabilization was achieved by addition to the formulation of a cationic surfactant, dodecylamine acetate, physical developers thus prepared being stable for months. Jonker and his co–workers [93] also investigated the possibility of preparing physical developers of non–noble metals such as copper and lead using suitable redox couples such as titanium and vanadium. Some of these were successful in photoplating but there has been no report of their ability to reveal latent fingerprints.

Goode and Morris [9] have reviewed early work on this technique from its first use to enhance fingerprints revealed by metal deposition. The authors further reported that stabilized physical developer reacts with lipid material present in the fingerprint desposits although the mechanism of reaction is not fully understood. Preparation of the reagent is described together with its use in revealing latent fingerprints on paper, patterned non–absorbent and absorbent surfaces and pressure sensitive tapes. Although there is little information in the literature, the reagent is undoubtedly very successful and is now extensively used by UK police forces. Because lipid material, with which the reagent reacts, is insoluble in water the technique is suitable for exhibits that have been exposed to wet or damp conditions. A further advantage is that it may also be successfully employed after ninhydrin treatment.

Enhancement Procedures

A radioactive visualisation method has been reported by Knowles [11] for cases where there is low contrast between the black fingerprint visualised by physical developer and the background surface. The silver image is first converted to silver bromide which is then reacted with radiolabelled thiourea to produce silver sulphide. Goode and Morris [9], in describing the method, have pointed out that in some cases printing ink will absorb radioactive sulphur 35. However this was not thought to be a serious limitation.

An alternative approach has been described by Nolan et al. [94] who successfully employed scanning electron microscopy to remove interference from backgrounds. The method allows fingerprint sized samples to be examined without cutting the paper on which the print resides.

2.4.2 Small Particle Reagent

Following on from research into the physical developer process Morris and Wells [95] patented a technique for developing fingerprints which involved suspending finely divided particles in a surfactant solution. A wide range of combinations of particulate materials (including many powdered metals) and surfactants were examined. The preferred formulation comprised molybdenum disulphide dispersed in Tergitol 7 and choline chloride. The method was later described in detail by Goode and Morris [9] while more recently Pounds and Jones [96] reported that choline chloride was not an essential component of the formulation. These authors recommended the use of molybdenum disulphide dispersed in Manoxol OT. The technique, which can be used as a spray at scenes of crime or as a dish development technique for laboratory use, is effective on a wide range of materials such as polythene, paint and glass etc. It is also capable of revealing fingerprints on wet or damp surfaces. The reagent appears to react with water insoluble unsaturated lipids such as oleic acid [9].

Enhancement Procedures

The procedure described by Nolan et al. [94] to enhance fingerprints revealed by physical developer is equally applicable to those produced by small particle reagent. Great care must, however, be taken in handling objects as the developed fingerprint can be easily smudged.

Future Work

Several areas of development are desirable, particularly for physical developer, including the requirement to accelerate the reaction and to simplify preparation of the reagent. Identification of the mechanism of reaction may enable greater sensitivity to be achieved. Current enhancement procedures are not totally satisfactory and other approaches such as the use of fluorescent procedures may be particularly useful for overcoming problems of background interference.

2.5 Iodine

The capacity of iodine vapour to reveal latent fingerprints is one of the oldest techniques known with its use being described by Forgeot [97] as early as 1891.

There are two possible mechanisms by which iodine vapour reveals latent fingerprints; absorption into the lipid material present in the fingerprint deposit or iodination of unsaturated lipids such as oleic acid. Olsen [2] reports that the process is based on absorption and additionally recommends iodine as the first action technique of choice for paper. Support for a physical rather than chemical process came from Almog et al. [98] who studied iodine fuming of paper samples.

It is well known that, after visualisation, there is a rapid fading of the fingerprints. Goode and Morris [9] assign this loss partly to re–sublimation and partly to chemical reaction with unsaturated lipids, resulting in the formation of colourless compounds.

2.5.1 Iodine Fuming Methods

Several methods, used over the years, for fuming exhibits with iodine have been described by Olsen [2]; samples may be exposed to the vapour in a cabinet, by use of a fuming pipe or from iodine liberated by a chemical reaction. An alternative method is to dust with ground iodine crystals. Two popular methods for use on absorbent or non-absorbent surfaces have been described by Goode and Morris [9]; an iodine cloud method and one based on use of a fuming pipe. Using these techniques samples are saturated with iodine vapour and then the excess iodine is allowed to desorb until optimum intensity is achieved.

There are, however, limitations to the iodine fuming technique. Almog et al. [98] have listed these as rapid fading of the impressions, lack of contrast causing problems with photography and failure to reveal prints greater than 3 days old. The authors suggest a solution to the ageing problem by constructing an apparatus which allowed the print to be treated simultaneously with iodine fumes and water vapour. Results from this apparatus showed that latent prints, up to 110 days old could be revealed on paper. It was found that contrast was considerably improved and, additionally, that the new technique frequently performed better than ninhydrin.

2.5.2 Fixing of Iodine Fumed Fingerprints

The fugitive fingerprint, revealed by exposure to iodine vapour, has to be fixed prior to photography. Early fixation methods have been described by Olsen [2] including the use of starch solution while Trowell [99] describes the use of p,p′ – tetramethyldiaminodiphenylmethane (tetrabase) which produces a green blue coloured fingerprint stable for several months. Mashiko and Ishizaki [100] have reported the use of 7,8 – benzoflavone which produces a dark purple coloured print. Cyclohexane was chosen as solvent as it does not cause ink to run and is therefore suitable for use on documents. Both of the above methods have been examined in detail by Goode and Morris [9].

An interesting development stemming from the 7,8 – benzoflavone method has been reported recently by Haque et al. [101] who dissolved iodine in the fixation formulation. It was found that fingerprints several weeks old could be developed on various porous surfaces and, compared to iodine fuming, the technique showed improved sensitivity.

2.5.3 Use of Iodine Fuming to Reveal Fingerprints on Skin

Shin and Argue [102], Adcock [103] and Gray [104] have all reported the visualisation of fingerprints on skin by fuming with iodine. Fumed prints were transferred to a metal plate of silver or tin, and revealed by exposure to a light source. Detection times ranged from up to 2 hours for living skin to 48 hours for cadavers cooled to 4 °C. Hammer [105] and Hammer and Lindner [106] have reviewed this technique and compared it with others for revealing fingerprints on skin i. e. X-ray methods [107-111], lasers [47, 69], magnetic powder and Dakty foil [112, 113] and Kromecote lift techniques [114]. The two latter techniques were preferred. An alternative approach has been reported by Feldman [115] and Feldman et al. [116] in which the first step involved fuming with iodine, followed by placing a plastic strip coated with leuco crystal violet over the fumed area. A dark purple fingerprint was revealed. In laboratory tests using this method, prints were developed up to 1 hour from placement. The method was reported as being equal to, if not better than, the silver plate transference method.

2.6 Dyes and Stains

The use of dyes and stains for revealing latent fingerprints has been known for many years. In 1907 Corin and Stokis [117] outlined a procedure employing Sudan Red III. More recently Mitsui [118] developed the use of Sudan Black B to reveal fingerprints on paper that had been soaked with water, a circumstance where ninhydrin would not have been effective. The major use of dyes and stains, however, has been to reveal fingerprints on adhesive tapes and to enhance fingerprints laid down in blood.

2.6.1 Development of Fingerprints on Adhesive Tapes

Ishiyama [119] has reported successful development of fingerprints on the adhesive side of cellophane tape by use of coomassie brilliant blue 250 although no details are given.

Arima [120] reported using aqueous solutions of crystal violet or Victoria Page Blue for PVC, cloth or cellophane tape. In the same paper Arima suggests the use of the fluorescent dye, mitephor, for coloured and black electrical tapes. It was, however, found that the fluorescent reagents viz. o-phthalaldehyde and 8-anilino-1-naphthalene sulphonic acid were equally effective. An alternative method for revealing fingerprints on black tape, described by Wilson and McCleod [121], involved dye staining followed by transference of the revealed fingerprint. In this method the tape was first treated with a solution of gentian violet, excess dye removed, and fingerprints transferred to the emulsion side of photographic paper by pressing with a low heat iron. A problem with this procedure was described by Truszkowski and Loninga [122] who found that the emulsion side of the photographic paper often sticks to the adhesive side of the tape and suggested a modified approach where the tape is dried before the transfer step. A detailed description of the gentian violet procedure, followed if necessary for dark coloured tapes by transference to photographic paper, has recently been made by Goode and Morris [9].

An alternative approach was taken by Taylor and Mankevich [123] who recently reported using a silver protein staining procedure to reveal fingerprints on tape. The method was said to be effective when used in conjunction with gentian violet.

2.6.2 Enhancement of Blood Fingerprints

A number of dyes have been reported over the years for enhancing fingerprints in blood, these include leucomalachite green [2], luminol [2, 124], benzidine [2, 125] and ninhydrin [2]. When using ninhydrin on cloth, Olsen [2] recommended first fixing the blood fingerprints before treatment with a solution of ethyl alcohol to prevent diffusion of ridges. In view of its carcinogenicity the use of benzidine cannot be recommended.

In an attempt to seek a far less hazardous alternative to benzidine Holland et al. [126] have suggested the use of 3,3′, 5,5′-tetramethylbenzidine (TMB) which appears to be similar in sensitivity [127]. More recently a study has been made by Lee [128] using TMB on both absorbent and non-absorbent surfaces and cadaver skin. The author reported that a combination of TMB and ninhydrin seemed to yield excellent results although details have yet to be published.

Recent work by Nutt [129] described the successful use of a spray solution of o-toluidine, hydrogen peroxide and glacial acetic acid to reveal a blood print on the back of a dead female. It should, however, be remembered that o-toluidine is carcinogenic.

The most comprehensive review of enhancement techniques has been published by Olsen [130] in a comparison of the sensitivities of amido black, Burnau's benzidine solution, benzidine glacial acetic acid, benzidine free base, leucomalachite green, luminol, o-toluidine, phenolphthalein, ninhydrin and TMB for blood deposited on paper and glass. Details of preparation and methods of use are described for each process. The author concludes that benzidine is the reagent most likely to provide the best results in visualizing impressions made in blood.

To overcome problems in photographing weak blood marks on coloured surfaces, Parkin and Hartley [130A] reported an enhancement technique based on transference of haem from the mark to agar gel incorporating leucomalachite green. Blue-green coloured prints are revealed which require immediate photography as rapid diffusion of the print occurs. Best results were obtained with flat smooth non-porous surfaces. The technique was also useful for prints laid down in other body fluids such as semen and saliva using appropriate dyes.

Future Work

Particular attention should be given to fluorescent dyes and transference techniques which would not only give increased sensitivity over visual methods but might also aid observation of fingerprints laid on coloured or patterned backgrounds.

2.7 Silver Reagents

2.7.1 Silver Nitrate

The use of silver nitrate to reveal latent fingerprints dates back to the work of Forgeot [97], in 1891. The technique was used extensively to reveal fingerprints on non–porous or paper samples but was largely superceded in 1954 by ninhydrin [19]. The effectiveness of silver nitrate results from its reaction with chlorides present in the eccrine sweat portion of a fingerprint deposit. The product of reaction, silver chloride, rapidly turns black on exposure to light.

Early work on the reagent has been reviewed by Goode and Morris [9] who also optimised the method for use on newsprint and untreated wood by employing a methanolic solution of the reagent.

The main problem with developing fingerprints by silver nitrate is the tendency to produce high background staining which results from silver being absorbed into the surface under treatment. A further difficulty is in controlling the photochemical development of a treated fingerprint. The problem of background interferences has been investigated in detail by Morris [131]. This author has patented a method in which a sample is first treated with a solution of silver ethylenediaminetetraacetic acid (EDTA) followed by a reduction step to reveal the fingerprint. Recent work by Goode and his colleague Morris [9] has led to the development of a modified silver nitrate formulation (MSN) which they recommended for developing marks up to a few weeks old on certain types of paper and untreated wood.

The main limitation to the use of these silver based methods is that they are not suitable for exhibits that have been stored or exposed to high humidities. Under these conditions chloride in the fingerprint deposit migrates. This is well illustrated by Goode and Morris [9] who showed that significant deterioration occurs after 15 days at 60 % relative humidity. Recently Reed [132] has reported using an aqueous solution of silver nitrate to reveal fingerprints on brass counterfeit coins when treatment with cyananoacrylate ester powder had proved unsuccessful.

2.7.2 Other Silver Reagents

Alternatives to silver nitrate have been investigated by Kerr et al. [133] who studied silver perchlorate, chromate, tetrafluoroborate, hexafluoroantimonate and a silver perchlorate/camphor reagent. Of these the silver perchlorate/camphor reagent was regarded by the authors as a chemical alternative to metal deposition [134] for revealing latent fingerprints on polythene. A surface is treated successively with camphor, silver perchlorate, photographic developer and physical developer.

2.8 Powders

Powders are widely used for revealing latent fingerprints on non–porous surfaces. The technique, which relies on mechanical adhesion of the powder to fingerprint deposits, is easy to apply and results are immediately apparent. The use of dusting powders dates back to early this century and since that time many substances and formulations have been investigated. Olsen [2] has described application techniques and lists seventeen preparations based mainly on metals, metal oxides and metal salts mixed with either gum or rosin. Goode and Morris [9] have also reviewed the technique and list preparative details of twenty–one typical powder formulations. The authors considered the most sensitive for general use to be aluminium flake and dusting with this powder is now well established in the UK. The powder is used in the paint industry and is manufactured by grinding a paste of atomised aluminium, stearic/palmitic acids and mineral oil in a ball mill [135]. During the process the particles acquire a flat plate-like structure and a surface layer of fatty acids which aids adhesion with fingerprint deposits.

In the method employed at scenes of crime throughout the UK, the powder is commonly applied by dusting likely areas for latent fingerprints with a brush. The most popular brush, called a Zephyr brush, comprises a mop of fibre glass filaments approximately 5 cm long. In the case of aluminium powder the technique is completed by lifting the revealed fingerprint with low tack adhesive tape and transferring to a clear acetate sheet where it may be conveniently photographed. The natural high reflectivity of aluminium powder assists the photographic process.

For surfaces such as polythene, where the adhesive nature of brushing might damage the latent fingerprint, MacDonell [136] prepared magnetic powders which could be gently applied by means of a magnetic wand. A range of commercially available magnetic powders, termed 'Magna Brush', are now available. An application where the Magna Brush was claimed to be more effective than other powder techniques has been outlined by Barron et al. [137]. These authors found the technique to be particularly successful on fingerprints which have been rejuvenated by a freeze/thaw method prior to dusting.

An alternative brushless method has been described by Thomas [138] which is based on an electrostatic technique. Although the method was not successfully developed at that time for operational use, it formed the basis for equipment used to reveal indented writing [139, 140]. It has however more recently been applied successfully in case work to the lifting of finger marks in dust. [140A].

The problem of revealing latent fingerprints on skin has also been attacked by the use of powders. Reichardt et al. [114] describe a method where the latent print is first lifted off the skin surface using Kromekote paper and then revealed by powdering with Magna Brush or black powder. Fingerprints were revealed up to 1.5 h after application of a latent print to living skin. More success has, however, been achieved where Magna Brush, applied directly to the skin of a cadaver, revealed a fingerprint in a murder case [141].

2.8.1 Fluorescent Powders

Experiments on modifying fingerprint powders with a fluorescent dye have been described by Thornton [142]. The most satisfactory formulation comprised conventional black fingerprint powder blended with coumarin–6. Dusting with Mars Red latent fingerprint powder and other fluorescent pigments blended with hydrocarbon rosin, has been studied by Menzel and Duff [143] to extend the application of laser techniques. The laser dyes coumarin–6 and the rhodamines suitably compounded with rosin were expected by the authors to be useful. This work was followed by a study from Menzel and Fox [144] who prepared several formulations combining various fluorescent dyes with magnetic powder. The procedure allowed the detection of latent fingerprints by laser even on luminescing surfaces. The study also showed that formulations with Nile Blue, oxazine, DODC or 3,3′–diethylthiotricarbocyanine iodide (DTTC) were potentially useful for scenes of crime work using a portable irradiation system based on a helium neon laser. Later Kerr et al. [84] prepared their own magnetic and non–magnetic powders by mixing fluorescein and rhodamine B with cornstarch, calcium sulphate and gum arabic. Excellent results were obtained for fresh fingerprints deposited on non–porous surfaces such as plastics, glass, metal etc. and for multicoloured and coloured glossy surfaces.

There is a need for powders as effective as aluminium flake but of different colours or exhibiting fluorescence which will allow maximum contrast to be achieved with surfaces of a particular colour.

2.9 Metal Deposition (MD)

The technique dates form 1969 when Theys et al. [145] revealed fingerprints on paper by vacuum evaporation of a mixture of zinc, antimony and copper. The first detailed review of the technique, however, was produced by Kent et al. [146] who described development work undertaken in various establishments and in an operational trial. The process is operated in a vacuum chamber by securing an exhibit above a pair of molybdenum boats, one containing gold and the other cadmium. The item under study is treated successively by vacuum evaporation of gold followed by cadmium. The method relies on gold penetrating or diffusing into the fingerprint material rather than into the background or vice versa. Fingerprints are revealed when cadmium nucleation occurs onto the gold coating. The process is particularly successful on polythene objects, kitchen ware and containers but not for heavily plasticized PVC. Later work by Abe [147] suggested that the method was capable of detecting fingerprints on white synthetic fabric or thin cotton. The author described the use of a cadmium or gold coating followed by cadmium deposition but found the most successful procedure to be the one involving the use of an initial gold coating.

Kent [134] has subsequently reported that due to its toxic hazard cadmium should be replaced by zinc for operational use. The technique is now well established in casework in the UK as a very successful method for revealing latent fingerprints on polythene surfaces.

2.10 Radioactive Techniques

These techniques are based on converting a latent fingerprint into a radioactive form, fingerprints being visualised by sandwiching the sample between photographic film. The process, called autoradiography, is attractive as it offers a potential way of removing highly patterned background which otherwise would interfere with the revealed fingerprint.

The first application of radioactive techniques in the field of fingerprint detection was described by Takeuchi et al. [148] who, in 1958, used stearic acid and formaldehyde labelled with carbon 14 on paper, metal foil and glass. The technique was extended to polythene and leather by Akerman [149] and Stverak et al. [150] using a solution of silver nitrate labelled with silver 110. A vapour phase method was developed by Goode et al. [151] which produced bromination of unsaturated components present in fingerprint deposits. No attempt was made to optimise reaction conditions but fingerprints were successfully revealed on paper. The valuable radioactive method, sulphur dioxide, was proposed by Grant et al. [152] in 1963 as a result of their work on atmospheric pollution. The same authors later [153] described a technique for use on paper but few experiments were performed. The method was developed further by Spedding [154] who found that it was possible to visualise latent fingerprints on finely woven fabrics. Work into the underlying principles of the technique by Knowles [11] showed that the uptake of sulphur dioxide was limited to several unsaturated compounds present in the latent fingerprint. In a detailed study of the method, Goode and Morris [9] have recommended its use for revealing latent fingerprints on black or dark coloured PVC pressure sensitive tapes and have described a unit for processing samples.

The same authors [9] have also developed a method employing radioactive iodine monochloride which reacts, like sulphur dioxide and bromine, with unsaturated components of fingerprint deposits. The method was applicable for paper or plastic surfaces but is only of use on fingerprints which are less than about one day old.

Of all the methods described only radioactive sulphur dioxide has proved successful on operational material.

2.11 Electronography

This technique was first introduced by Grayam and Gray [107] in 1966 and involves bombarding lead dusted fingerprints with x–rays from high energy sources and detecting the emissions by photographic emulsions. Using this technique interferences from bachgrounds were virtually eliminated with success for multicoloured surfaces reported by Winstanley [109].

Potentially the most valuable application of the technique was in developing latent fingerprints on skin. Graham [108] in 1969 recorded fingerprints for up to 48 hours after application on amputated limbs, while Mooney [110] has reported developing good identifiable latent prints placed on a homicide victim. No operational success has, however, been reported for this technique.

2.12 Dimethylaminocinnamaldehyde (DMAC)

This reagent reacts with the urea content of latent fingerprints and was first investigated by Morris et al. [155] in 1973 and later patented [156]. Laboratory trials showed it to have benefits over ninhydrin as fingerprints were revealed within minutes after treatment. In addition the reagent appeared to be useful where reactions of surfaces with ninhydrin caused heavy background staining. Using the same formulation as that quoted in the patent, Sasson and Almog [157] examined the scope and limitations of the reagent and found that stains from latent fingerprints over 72 hours old were unresolved. In view of their results these authors concluded that the reagent would only be preferable to ninhydrin for rapid development of relatively fresh latent fingerprints when the application of heat for development was undesirable. Goode and Morris [9] have recently reviewed the reagent and make recommendations similar to those of Sasson and Almog.

2.13 Biological Detection Methods

2.13.2 Antisera and Lectins

Ishiyama et al. [158] and Okada and Ohrui [159] have reported the use of antisera and lectins to detect the presence of ABH blood group substances in latent fingerprint deposits. Although these reports concentrated on determining blood group types they also demonstrated that the technique was capable of revealing fingerprints. The method used was based on an agglutination reaction involving a two step incubation, firstly with anti–A, anti–B agglutinins or *Ulex europeus* anti–H lectin followed by appropriate red cell suspensions.

Following from this work, Hussain and Pounds [160], using a similar technique to Ishiyama et al. [158] evaluated anti–A, anti–B, various monoclonal antisera and a wide range of lectins, specifically for their ability to reveal latent fingerprints. Monoclonal anti–H and several lectins proved successful in revealing fingerprints on non–porous surfaces such as polythene and PVC, irrespective of blood group or secretor status of the donor.

2.13.2 Bacteriological Techniques

O'Neill [161] in 1941 demonstrated that it was possible to grow bacteria on fingerprints deposited on nutrient agar, but the technique had no practical use. Recently, Harper et al. [162] investigated the potential of bacteriological techniques by isolating organisms that are capable of growing on compounds present in sebum. Fingerprints were deposited on polythene, and agar gel incorporating the micro–organism laid on top. Following incubation the gel was removed and prints revealed on it by staining with naphthalene black 12B. Of 300 organisms investigated, *Acinetobacter calcoaceticus* proved best.

Future Work

Little research has at present been undertaken into the use of biological techniques. Methods involving antibodies, enzyme and micro–organisms etc., routinely employed in biochemical laboratories, are renowned for their great sensitivity. Therefore, research in this field must surely offer us, for the future, techniques of greater sensitivity than at present.

3 References

1. Gill P, Jeffreys A J, Werrett D J (1985) Nature *318*:577
2. Olsen R D (1978) in: Scotts Fingerprint Mechanics. Charles C Thomas, Springfield, Illinois, USA
3. Moenssens A A (1975) in: Fingerprint Techniques. Chilton Book Company, Pennsylvania, USA
4. Cowger J F (1983) in: Friction Ridge Skin. Elsevier, Oxford
5. Weiner J S, Hellmann K (1960) Biol. Rev. *35*(2):141
6. Montagna W (1962) in: Advances in Biology of Skin. Vol 111 Eccrine Glands and Eccrine Sweating, Montagna W, Ellis R A, Silver A F, (eds), Pergamon Press, London
7. Woodroffe R C S, Shaw D A (1974) in: The Normal Microbial Flora of Man: Natural Control and Ecology of Microbial Populations on Skin and Hair. Skinner F A, Carr J G, (eds) Academic Press, London
8. Kuno Y (1956) in: Human Perspiration. Charles C Thomas, Springfield, Illinois, USA
9. Goode G C, Morris J R (1983) in: Latent Fingerprints: A Review of Their Origin, Composition and Methods of Detection, Atomic Weapons Research Establishment, Aldermaston, Reading, Berkshire, Report No. 022/83
10. Manual of Fingerprint Development Techniques (1986) Home Office, Scientific Research and Development Branch, London
11. Knowles A M (1978) J. Phys. E: Sci. Instrum. *11*:713
12. Lewis C A, Hayward B J (1971) in: Modern Trends in Dermatology, Human Skin Surface Lipids, Borrie P (ed) Butterworth, London
13. Powe W C (1972) in: Detergency, Theory and Test Methods, Dekker, New York, USA
14. Kellum R E (1967) Arch. Derm. *95*:218
15. Marples R R, Downing D T, Kligman A M (1971) J. Invest. Derm. *56*(2):127
16. Ruhemann S (1910) J. Chem. Soc. *97*:1438
17. Ruhemann S (1910) J. Chem. Soc. *97*:2025
18. Bottom C B, Hana S S, Siehr D J (1978) Biochem. Ed. *6*:4
19. Oden S, von Hofsten B (1954) Nature *173*:449
20. Oden S (1957) British Patent No. 767 341
21. Moore S, Stein W H (1948) J. Biol. Chem. *176*:367
22. Lamothe P J, McCormick P G (1972) Anal. Chem. *44*(4):821
23. Rispling O (1971) Int. Crim. Pol. Rev. *245*:30
24. Linde H G (1975) J. Forens. Sci. *20*:581
25. Crown D A (1969) J. Crim. Law Criminol. Pol. Sci. *60*(2):258
26. Morris J R, Goode G C (1974) Pol. Res. Bull. (24):45
27. Tighe D J (1984) Ident. News *34*(6):3
28. Stoilovic M, Warrener R N, Kobus H J (1984) Forens. Sci. Int. *24*:279
29. Herod D W, Menzel E R (1982) J. Forens. Sci. *27*(1):200
30. German E R (1981) Ident. News *31*(9):3
31. Creer K E (1983) Forens. Sci. Int. *23*:149
32. Seitz W R (1980) in: Fluorescent Derivatisation, CRE Critical Reviews in Analytical Chemistry
33. Herod D W, Menzel E R (1982) J. Forens. Sci. *27*(3):513
34. Menzel E R (1982) Ident. News *32*(2):3
35. Kobus H J, Stoilovic M, Warrener R N (1983) Forens. Sci. Int. *22*:161
36. Kobus H J, Stoilovic M, Warrener R N (1984) J. Forens. Sci. Soc. *24*(4):405
37. Warrener R N, Stoilovic M, Kobus H J (1984) J. Forens. Sci. Soc. *24*(4):405

37A. Stoilovich M, Kobus H J, Margot P A, Warrener R N (1986) J. Forensic Sci. *31*(2):432

38. Menzel E R, Everse J, Everse K E, Sinor T W, Burt J A (1984) J. Forens. Sci. *29*(1):99
38A. Everse K E, Menzel E R (1986) J. Forens. Sci. *31*(2):446
39. Almog J, Hirshfeld A, Klug J T (1982) J. Forens. Sci. *27*(4):912
40. Becker H D (1964) J. Org. Chem. *29*(2):1358
41. Jones D W, Wife R L (1972) J. Chem. Soc. Perkin 1, *3*:2722
42. Meier R, Lotter H G (1957) Chem. Ber. *90*(1):222
43. Menzel E R, Almog J (1985) J. Forens. Sci. *30*(2):371
43A. Lennard C J, Margot P A, Stoilovich M, Warrener R N (1986) J. Forensic Sci. Soc. *26*(5):323
44. Fritz H, Jordan H (1970) Archiv für Kriminologie *7*:163
45. Menzel E R (1980) in: Fingerprint Detection with Lasers, Marcell Dekker, New York, USA
46. Menzel E R (1985) J. Forens. Sci. *30*(2):383
47. Dalrymple B E, Duff J M, Menzel E R (1977) J. Forens. Sci. *22*:106
48. Duff J M, Menzel E R (1978) J. Forens. Sci. *23*(1):129
48A. Dikshitube Y S, Prasad L, Paul J N, Rao C V N (1986) Forensic Sci. Int. *31*:261
49. Salares V R, Eves C R, Carey P R (1979) Forens. Sci. Int. *14*:229
50. Dalrymple B E (1982) J. Forens. Sci. *27*(4):801
51. Matharu S S, Russell J R, Creer K E (1983) Forens. Sci. Int. *21*:197
52. Zahradnik M (1982) in: The Production and Applications of Fluorescent Brightening Agents, Wiley & Sons, Chichester
53. Ghosh P B, Whitehouse M W (1968) Biochem. J. *108*:155
54. Warrener R N, Kobus H J, Stoilovic M (1983) Forens. Sci. Int. *23*:179
54A. Kent T (1985) Paper presented at IDENTA-85, Jerusalem, Israel
54B. Watkin J E, Carey P R (1981) National Research Council, Canada Report LTR-PSP5
54C. Creer K Personal Communication
55. Kessel D, Belton J G (1975) Cancer Res. *35*:3735
56. Nelson J O, Warren P F (1981) Mutat. Res. *88*:351
57. Weber G (1952) Biochem. J. *51*:155
58. Gray W R, Hartley B S (1963) Biochem. J. *89*:59
59. Seiler N (1970) Methods Biochem. Anal. *18*:259
60. Burt J A, Menzel E R (1985) J. Forens. Sci. *30*(2):364
61. Udenfriend S, Stein S, Bohlen P, Dairman W, Leimgruber W, Weigele M (1972) Science *178*:871
62. Felix A M, Jimenez M H (1974) J. Chromatogr. *89*:361
63. Touchstone J C, Sharma J, Dobbins M F, Hansen G R (1976) J. Chromatogr. *124*:111
64. Roth M (1971) Anal. Chem. *43*(7):880
65. Simons S S, Johnson D F (1977) Anal. Biochem. *82*:255
66. Benson J R, Hare P E (1975) Proc. Nat. Acad. Sci. *72*(2):619
67. Mayer S W, Meilleur C P, Jones P F (1978) J. Forens. Sci. Soc. *18*:233
68. Almog J, Gabay A (1980) J. Forens. Sci. *25*(2):408
69. Menzel E R (1982) J. Forens. Sci. *27*(4):918
70. Fischer J F (1984) Ident. News *34*(7):2
71. Lee H C, Gaensslen R E (1984) Ident. News *34*(6):8
72. Lee K (1982) Can. Med. Assoc. J. *127*:359
73. Reed F A (1980) Pol. Res. Bull. No. 35/36, 32
74. Kendall F G (1982) Ident. News *32*(2):3
75. Kendall F G, Rehn B W (1983) J. Forens. Sci. *28*(3):777
76. Martindale W E (1983) Ident. News *33*(11):13
77. Olnik J H (1984) J. Forens. Sci. *29*(3):881
78. Besonen J A (1983) Ident. News *33*(2):3
79. Woods D D (1983) Ident. News *33*(4):10
80. Mock J P (1985) Ident. News *35*(9):7
81. Sahs P T, Wojcik R J (1984) Ident. News *34*(9):9
82. Gilman P L, Sahs P T, Gorajczyk J S (1985) Ident. News *35*(3):7
83. Illsley C P (1984) Ident. News *34*(1):6
84. Kerr F M, Barron I W, Haque F, Westland A D (1983) Can. Soc. Forens. Sci. J. *16*(1):39
85. Kobus H J, Warrener R N, Stoilovic M (1983) Forens. Sci. Int. *23*:233

86. Menzel E R, Burt J A, Sinor T W, Tubach-Ley W B, Jordan K J (1983) Forens. Sci. *28*(2):307
87. Vaughn J M (1985) Ident. News *35*(1):3
88. Menzel E R (1985) J. Forens. Sci. *30*(2):383
89. Arnold R R, Gallant J R (1985) Ident. News *35*(7):12
90. Jonker H, Dippel C J, Houtman H J, Janssen C J G F, van Beek L K H (1969) Photog. Sci. Eng. *13*(1):1
91. Mees C E K, James T H (1966) in: The Theory of the Photographic Process, Collier-MacMillan, London
92. Jonker H, Molenaar A, Dippel C J (1969) Photog. Sci. Eng. *13*(2):38
93. Jonker H, van Beek L K H, Dippel C J, Janssen C J G F, Molenaar A, Spiertz E J (1971) J. Photog. Sci. *19*:96
94. Nolan P J, Brennan J S, Keely R H, Pounds C A (1984) J. Forens. Sci. Soc. *24*:419
95. Morris J R, Wells J M (1979) British Patent No. 1 540 147
96. Pounds C A, Jones R J (1983) Trends Anal. Chem. *2*:180
97. Forgeot R (1891) Arch. d'Anthropol. Crim. *6*:387
98. Almog J, Sasson Y, Anati A (1979) J. Forens. Sci. *24*(2):431
99. Trowell F (1975) J. Forens. Sci. Soc. *15*:189
100. Mashiko K, Ishizaki M (1977) Ident. News, *27*(11):3
101. Haque F, Westland A, Kerr F M (1983) Forens. Sci. Int. *21*:79
102. Shin D H, Argue D G (1976) Can. Soc. Forens. Sci. J. *9*(2):81
103. Adcock J M (1977) J. Forens. Sci. *22*:599
104. Gray C (1978) J. Forens. Sci. Soc. *18*:47
105. Hammer H-J (1980) Forens. Sci. Int. *16*:35
106. Hammer H-J, Lindner R (1980) Z. Arztl. Fortbild. *74*(8):378
107. Grayham D, Gray H C (1966) J. Forens. Sci. *11*(2):124
108. Grayham D (1969) J. Forens. Sci. *14*(1):1
109. Winstanley R (1977) J. Forens. Sci. Soc. *17*:121
110. Mooney D J (1977) Ident. News *27*(2):5
111. Weiss Von K G, Perry R (1972) Beitr. Gerichtl. Med. *29*:223
112. Lindner R, Hammer H-J (1977) Kriminol. Forens. Wiss. H. *30*:63
113. Hammer H-J, Lindner R (1979) Kriminol. Forens. Wiss. H *36*:115
114. Reichardt G J, Carr J C, Stone E G (1978) J. Forens, Sci. *23*(1):135
115. Feldman M A (1981) in: New Applications of Chemistry to Criminalistics, A. Recovering Fingerprints from Skin, PhD Thesis, Kansas State University, USA
116. Feldman M A, Meloan C E, Lambert J L (1982) J. Forens. Sci. *27*(4):806
117. Corin G, Stokis E (1907) Ann. Soc. Med. Leg. Belg. *18*(1):17
118. Mitsui T, Katho H, Shimada K, Wakasugi Y (1980) Ident. News *30*(8):9
119. Ishiyama I (1981) J. Forens. Sci. *26*(3):570
120. Arima T (1981) Ident. News *31*(2):9
121. Wilson B L, McLeod V D (1982) Ident. News *32*(3):3
122. Truszkowski G, Laninga K (1984) Ident. News *34*(2):2
123. Taylor E M, Mankevich A (1984) Ident. News *34*(7):4
124. Zweidinger R A, Lytle L T, Pitt C G (1973) J. Forens. Sci. *18*(4):296
125. Harrell J R (1973) Criminol. *8*:49
126. Holland V R, Saunders B C, Rose F L, Walpole A L (1974) Tetrahedron *30*:3299
127. Garner D D, Cano K M, Peimer R S, Yeshion T E (1976) J. Forens. Sci. *21*(4):816
128. Lee H C (1984) Ident. News *34*(3):10
129. Nutt J (1983) Ident. News *33*(10):10
130. Olsen R D (1985) Ident. News *35*(8):10
130A. Parkin B H, Hartley K P (1987) Forensic Sci Int.
131. Morris J R (1974) British Patent No. 1 497 791
132. Reed R (1985) Ident. News *35*(7):11
133. Kerr F M, Westland A D, Haque F (1981) Forens. Sci. Int. *18*:209
134. Kent T (1981) J. Forens. Sci. Soc. *21*:15
135. Shackleton R, Wendon G (1972) Pig. Resin Tech. Sept. 27
136. MacDonell H L (1974) Ident. News *11*(3):11
137. Barron I W, Haque F, Westland A D (1981) Ident. News *31*(11):14

138. Thomas G L (1978) J. Phys. E: Sci. Instrum. *11*:722
139. Foster D J, Morantz D J (1979) Forens. Sci. Int. *13*:51
140. Ellen D, Foster D J, Morantz D J (1980) Forens. Sci. Int. *15*:53
140A. Brennan J (Personal communication)
141. Haslett M (1983) Ident. News *33*(2):7
142. Thornton J I (1978) J. Forens. Sci. *23*(3):536
143. Menzel E R, Duff J M (1979) J. Forens. Sci. *24*(1):96
144. Menzel E R, Fox K E (1980) J. Forens. Sci. *25*(1):150
145. Theys P, Yurgis Y, Lepareux A, Chevet G, Ceccaldi P F (1968) Int. Crim. Pol. Res. 217:106
146. Kent T, Thomas G L, Reynoldson T E, East H W (1976) J. Forens. Sci. Soc. *16*:93
147. Abe S (1978) Kag. Kei. Ken. Hok. *31*(4):377
148. Takeuchi T, Sakaguchi M, Nakamoto Y (1958) Naturw. *45*:36
149. Akerman K (1966) Int. J. Appl. Radiat. Isotop. *17*:657
150. Stverak B, Kopejtko J, Chodora F, Chyska J (1974) Radioisotopy *15*(6):805
151. Goode G C, Morris J R, Wells J M (1979) J. Radioanal. Chem. *48*:17
152. Grant R L, Lyth Hudson F, Hockey J A (1963) Nature *200*:1348
153. Grant R L, Lyth Hudson F, Hockey J A (1963) J. Forens. Sci. Soc. *4*:85
154. Spedding D J (1971) Natura *229*:123
155. Morris J R, Goode G C, Godsell J W (1973) Pol. Res. Bull. *21*:31
156. Morris J R (1976) British Patent No. 1 428 025
157. Sasson Y, Almog J (1978) J. Forens. Sci. *23*(4):852
158. Ishiyama I, Orui M, Ogawa K, Kimura T (1977) J. Forens. Sci. *22*(2):365
159. Okada T, Ohrui M (1978) Act. Crim. Japan *44*(3):94
160. Hussain J I, Pounds C A (1984) J. Forens. Sci. Soc. *24*(4):386
161. O'Neill M E (1941) J. Crim. Law Criminol. Pol. Sci. *32*:482
162. Harpter D R, Clare C M, Heape C D, Brennan J, Hussain J I (1987) Forens. Sci. Int. *33*(3):209

Subject Index

Affinity constant see Equilibrium constant
Antibodies, monoclonal 50–51
Antigen-antibody saturation curve 10
Antisera 115
—, anti-A, anti-B agglutinins 115
— monoclonal 115
Antiserum, collection 49–50
—, control of specificity 34–35
—, enantiomer recognition 35
—, isolation of IgG fraction 50
—, mixed 23, 24
—, polyclonal 10
— production 34–51
— storage 49–50
— titre 18, 20

Bacteria 115, 116
— dyes 115
Binding, non-specific 18, 19
Bolton-Hunter reagent 32
Bridge recognition 31–32

Cyanoacrylate enhancement 105
— — dyes 105
— — luminescent dyes 105
— — powders 105
— ester 104–106
— firearm examination 106
— hydrogen cyanide 105
— polymerisation mechanism 104
— vapour evolution 105
— — heat 105
— — humidity 105

Dimethylaminocinnamaldehyde (DMAC) 115
—, print diffusion 115
Disequilibrium assay 21
Drugs, radioimmunassay of
— analgesics and narcotics 58
— antiasthmatics 58
— antibiotics and antineoplastics 59
— anticholinergic drugs 59
— anticoagulants 59
— anticonvulsants 60
— antidiabetic drugs 60
— antidiarrhoeal drugs 60
— antiemetic drugs 60
— antihistamines 60
— anti-inflammatory drugs and gout suppressants 60–61
— antimalarial drugs 61
— antipsychotic, hypnotic and sedative drugs 61
— — barbiturates 61
— — benzodiazepines 61
— — phenothiazines 61
— — tricyclic antidepressants 62
— — other antipsychotic drugs 62
— antithyroid drugs 62
— antitubercular drugs 62
— anti-ulcer drugs 62
— antiviral drugs 62
— anti-worm (anthelmintic) drugs 62
— cardiovascular drugs 63
— — adrenergic blocking drugs 63
— — antiarryhmic drugs 63
— — antihypertensive drugs 63
— — cardiac depressants 63
— — — glycosides and aglycones 63–64
— CNS stimulants and psychoactive drugs 64
— diagnostic aids 64
— diuretic drugs 64
— eicosanoids, leukotrienes and prostaglandins 64
— ergot derivatives 64–65
— ganglionic stimulant drugs 65
— herbicides and insecticides 65
— immunomodulating and immunosuppressive drugs 65
— muscle relaxants 65
— narcotic antagonists 65
— renal tubular transport inhibitors 65
— steroids 66
— sympathomimetic drugs 66
— tobacco alkaloids and metabolites 66
— toxins 66–67
— vasodilators 67
— vitamins 67

Dyes 109–111
—, adhesive tape 109f
— — transference 110
—, bacteria 115
—, blood fingerprints 110f
—, cyanoacrylate ester enhancement 105f
—, — luminescent 105f
—, iodine 109
—, — skin 109

Electronography 114
— skin 114
Emit 34, 48
β-Emitting isotopes 25
— ^{14}C 25
— energy spectrum 26
— ^{3}H 25
— liquid scintillation counting 26–28
— — dual isotope counting 27
— — quenching and quench correction 27–28
— pulse amplitude spectrum 26–27
— ^{35}S 25
γ-Emitting isotopes 29
— ^{57}Co 29
— γ-counting 33
— ^{123}I 29
— ^{125}I 29
— ^{131}I 29
— ^{113m}In 29
— ^{75}Se 29
Endogenous compounds in samples 15, 16
Equilibrium cónstant 8

Fingerprint, apocine gland 94
—, eccrine gland 93
— deposit composition 93f
—, sabaceous gland 94
Freund's complete and incomplete adjuvants 49

Hapten 34
— bifunctional reagents 32, 37
— reactive groups 38–39
Heterogeneous assays 7
Heterologous assays 31–32
Homogeneous assays 7
Homologous assays 31–32

Immunization 49
Immunogen 34
— bridge length 35
— degree of conjugation 35, 47–48
— non-specifically bound hapten 48
— preparation 34–47
— — from active hydrogen 44–45
— — — — aromatic 40–41
— — — — primary aliphatic 36–37
— — — — secondary aliphatic 37–40
— — — carbonyl groups 44
— — — carboxyl groups 41–42
— — — ester linkages 45
— — — hydroxyl aliphatic 42–43
— — — — phenolic 43
— — — lactone rings 45–46
— — — nitro groups 41
— — — thiol groups 46
— — from miscellaneous haptens 46–47
— purification 47
— storage 48
Iodine 108f
—, fixing 109
—, fuming 108
—, mechanism of reaction 108
—, skin 109
—, — dyes 109
—, — transference 109

Late addition assay 21
LIDIA 55
Liquid scintillant 26
Lectins 115
Luminescence 98–106, 113
—, cyanoacrylate dyes 105f
—, — ester enhancement 105f
—, fingerprint age 100
—, — contamination 100
—, latent fingerprints 100
—, narrow band filters 101
—, ninhydrin anaologues 99f
—, portable light source 104
—, powders 113
—, Rhuemann's purple 97f
Luminescent derivatives 101–114
—, 1-dimethylaminonaphthalene-5-sulphonyl chloride (Dansyl chloride) 102
—, 4-phenylspino(furan-2-(3H), 1'-phthalan-3,3' dione (Fluorescamine) 102f
—, 7-chloro-4-nitrobenzo-2-oxa-1,3-diazole (NBD-CR) 101f
—, 0-phthalaldehyde 103
—, vapour phase reagents 103f

Metal deposition (MD) 113
— mechanism 113
— process details 113
Mixed antisera 23, 24
Monoclonal antibodies 50–51
Multiple drug assay 24

Ninhydrin 95–100
— analogues 99f
—, enhancement see Rhuemann's purple 97f
—, enzymes pre treatment 98f
—, heat 96f
—, humindity 96f
—, mechanism amino acids 95
—, metal complexation 99f
—, — luminescence 99f
—, non-polar solvents 96f
Non-specific binding 18, 19

PALA 55
Physical developer 106f
— enhancement 107
— mechanism 106f
—, — radioactive silver sulphite 107
—, — scanning electron microscopy 107
Polyclonal antiserum 10
Powders 105, 112f
—, cyanoacrylate ester enhancement 105
—, flake manufacture 112
—, luminescent 113
—, magnetic 112
—, mechanism adhesion 112
—, scenes of crime 112

Radioactivity 107, 114
—, carbon 14 label 114
—, iodine monochloride 114
—, physical developer silver sulphide 107
—, silver nitrate 114
—, sulphur dioxide 114
Radioligands 25–33
— β-labelled drugs 25
— — stability 25–26
— — storage 25–26
— γ-labelled drugs 29–33
— — stability 32–33
— — storage 32–33
— specific activity 33
Radioiodination methods 29–33
— using 29f
— — N-bromosuccinimide 30
— — chloramine T 29–30
— — conjugation labelling with prosthetic groups 31–32
— — electrolysis 30
— — enzymatic methods 30–31
— — heating in melt 32
— — iodine monochloride 31
— — Iodobeads 30
— — Iodogen 30
— — molecular iodine 31
— — oxidative methods 29–31
— — Protag-125 30
— — recoil labelling 32
— — substitution methods 32
Radioimmunoassay
— accuracy 16
— antiserum dilution curve 17–19, 21
— basic principles 7–8
— calibration curve 7–8, 19–23
— — heteroscedasticity 15
— cross-reactivity 22
— data processing 8
— detection limit 15
— general assay 17, 19, 23–24
— hazards 57
— metabolites 16, 17, 22
— optimisation of assay conditions 11–14, 20–21
— practical design of an assay 17–24
— precision 14–15
— precision profile 15
— quality assurance 55–56
— sensitivity 15–16
— separation of bound and free fractions 7, 28, 51–55
— — using 7f
— — — adsorption 51–52
— — — dialysis 55
— — — electrophoresis 54
— — — fractional precipitation 52–53
— — — gel permeation chromatography 55
— — — ion exchange 55
— — — partition between two liquid phases 55
— — — second antibody precipitation 53–54
— — — solid phases 54
— specific assay 17, 19, 20–23
— theory 7–16
— — antigen-antibody equilibrium 7
— — Ekins' conditions for sensitive assay 13
— — Ekins' equation 12
— — insensitive assay for high analyte concentrations 14
— — Scatchard plot 8–11
— — theoretical immunoassay model 11–14
— — Yalow and Berson's conditions for sensitive assay 13
Reverse pipetting 57
Rhuemann's luminescence 97f
— metal complexation 98
— — heat 98
— — humidity 98
— — luminescence 98
— purple 97f
RIA see Radioimmunoassay

Silver, ethylenediaminetetra acetic acid 111
— nitrate 111, 114

—, other silver salts 111 f
—, print diffusion 111
—, radioactive 114
Small enhancement 108
— — scanning electron microscopy 107
— particle reagent 107 f

Wood's reagent 32

Author Index Volumes 1 – 3

The volume numbers are printed in italics

Hellmiss, G.: Thermal Analysis Methods in Forensic Science. *2*, 1–30 (1988).

Mathyer, J.: Optical Methods in the Examination of Questioned Documents. *2*, 31–46 (1988).

Mukoyama, H. and Seta, S.: The Determination of Blood Groups in Tissue Samples. *1*, 37–90 (1986).

Pounds, C. A.: Developments in Fingerprints Visualisation, *3*, 91–119 (1988).

Riederer, J.: The Detection of Art Forgeries with Scientific Methods. *1*, 153–168 (1986).

Sellier, K.: Death: Accident or Suicide by Use of Firearms. *1*, 91–116 (1986).

Seta, S., Sato, H., Miyake, B.: Forensic Hair Investigation. *2*, 47–166 (1988).

Smith, R. N.: Radioimmunoassay of Drugs in Body Fluids in a Forensic Context. *3*, 1–89 (1988).

Thatcher, P. J.: The Scientific Investigation of Fire Causes. *1*, 117–152 (1986).

Thornton, J. I.: Forensic Soil Characterization. *1*, 1–36 (1986).